Albert Philson Brubaker

A Compend of Human Physiology

Albert Philson Brubaker

A Compend of Human Physiology

ISBN/EAN: 9783337365653

Printed in Europe, USA, Canada, Australia, Japan

Cover: Foto ©berggeist007 / pixelio.de

More available books at **www.hansebooks.com**

? QUIZ-COMPENDS. ? No. 4.

A

COMPEND

OF

HUMAN PHYSIOLOGY.

ESPECIALLY ADAPTED FOR THE USE OF
MEDICAL STUDENTS.

BY

ALBERT P. BRUBAKER, A.M., M.D.,

DEMONSTRATOR OF PHYSIOLOGY IN THE JEFFERSON MEDICAL COLLEGE; PROFESSOR
OF PHYSIOLOGY, PENNSYLVANIA COLLEGE OF DENTAL SURGERY;
MEMBER OF THE PATHOLOGICAL SOCIETY
OF PHILADELPHIA.

FOURTH EDITION, REVISED AND ENLARGED.

WITH ILLUSTRATIONS

AND

A TABLE OF PHYSIOLOGICAL CONSTANTS.

PHILADELPHIA:

P. BLAKISTON, SON & CO.,

1012 WALNUT STREET.

1888.

PRESS OF WM. F. FELL & CO.,
1220-24 Sansom Street.

PREFACE TO FOURTH EDITION.

A fourth edition of the Compend of Physiology having been called for, the author has taken the opportunity to revise many of the sections, to insert a few figures, and to add some ten pages of additional matter, which, it is hoped, will make the book more complete and increase its usefulness to the student.

1210 Race Street.   A. P. BRUBAKER.

PREFACE TO SECOND EDITION.

This Compend of Physiology is the outgrowth of the author's system of examinations in the Quiz room during a number of years, and was written at the request of medical students who desired a compact and convenient arrangement of the fundamental facts of human physiology. As most medical students enter upon the study of physiology before they have acquired a thorough knowledge of anatomy, it was thought desirable that such anatomical details should also be inserted as would be essential to a clear conception of the functions about to be studied. It was believed that it would be practically useful to students during their attendance upon lectures and in reviewing the subject prior to examinations. The fact that during the first year after its publication the first edition has been exhausted, proves that it has met the needs of students.

In preparing a second edition the author has carefully revised the entire work, and inserted some fifteen pages of additional matter, which it is hoped will still further increase the usefulness of the book.

To those teachers of physiology who have kindly noticed and recommended the Compend to their students I tender my thanks, and trust that in its improved condition it will continue to merit their approval.

ALBERT P. BRUBAKER.

TO MY FATHER,

HENRY BRUBAKER, A.M., M.D.,

THIS LITTLE VOLUME

IS AFFECTIONATELY INSCRIBED.

TABLE OF CONTENTS.

COMPEND

OF

HUMAN PHYSIOLOGY.

Physiology, from φύσις, nature, and λόγος, a discourse, in its original application embraced the study of all natural objects, inorganic as well as organic. In its modern application physiology signifies *the study of life;* the investigation of the vital phenomena exhibited by all organic bodies, vegetable and animal.

It may be divided into—

1. *Vegetable physiology,* which treats of the phenomena manifested by the several structures of which the plant is composed.

2. *Animal physiology,* which treats of the phenomena manifested by the organs and tissues of which the animal body is composed.

Human Physiology is the study of the functions exhibited by the human body in a state of health.

A *function* is the action of an organ or tissue.

The Functions of the Human Body may be classified into three groups, viz.:—

1. *Nutritive functions,* which have for their object the preservation of the individual; *e. g.,* digestion, absorption, circulation of the blood, respiration, assimilation, animal heat, secretion and excretion.

2. *Animal functions,* which bring the individual into conscious relationship with external nature; *e. g.,* sensation, motion, language, mental and moral manifestations.

3. *Reproductive function,* which has for its object the preservation of the species.

The facts of human physiology have been determined by means of anatomy, chemistry, pathology, comparative anatomy, vivisection, the application of physics, etc.

The body may be studied from a *chemical* and *structural* point of view.

B   9

CHEMICAL COMPOSITION OF THE HUMAN BODY.

Of the Sixty-four Chemical Elements, about *sixteen* enter into the composition of the body, in the following proportions :—

Oxygen..........72.00 ⎤ O. H. and C. are found in all the tissues and
Hydrogen...... 9.10 ⎱ fluids of the body, without exception.
Nitrogen......... 2.50 ⎰ O. H. C. and N. found in most of the fluids
Carbon..........13.50 ⎦ and all tissues except fat.

Sulphur.......... .147 ......In fibrin, casein, albumen, gelatin; as potassium-sulpho-cyanide in saliva; as alkaline sulphate in urine and sweat.

Phosphorus... 1.15 ......In fibrin and albumen; in brain; as tri-sodium phosphate in blood and saliva, etc.

Calcium........ 1.30 ......As calcium phosphate in lymph, chyle, blood, saliva, bones and teeth.

Sodium......... .10 ......As sodium chloride in all fluids and solids of the body, except enamel; as sodium sulphate and phosphate in blood and muscles.

Potassium...... .026 ......As potassium chloride in muscles; generally found with sodium as sulphates and phosphates.

Magnesium.... .001 ......Generally in association with calcium, as phosphate, in bones.

Chlorine.. ..... .085. ......In combination with sodium, potassium and other bases, in all the fluids and solids.

Fluorine........ .08 ......As calcium fluoride in bones, teeth and urine.
Iron...... ...... .01 ......In blood globules; as peroxide in muscles.
Silicon..........a trace ......In blood, bones and hair.
Manganesium a trace ......Probably in hair, bones and nails.

Of the four chief elements which together make up *97 per cent.* of the body, O. H. N. are eminently *mobile, elastic,* and possess great *atomic heat.* C. H. N. are distinguished for the narrow range and feebleness of their affinities and chemical inertia. C. has the greatest atomic cohesion. O. is noted for the number and intensity of its combinations, and its remarkable display of chemical activity.

Chemical elements do not exist alone in the body, but are combined in characteristic proportions to form compounds, the *proximate principles,* which are the ultimate compounds to which the fluids and solids can be reduced.

Proximate Principles exist in the body under their own form, and can be extracted without losing their distinctive properties.

There are about *one hundred* proximate principles, which are divided into four classes, viz.: *inorganic, organic non-nitrogenized, organic nitrogenized,* and *principles of waste.*

I. INORGANIC PROXIMATE PRINCIPLES.

SUBSTANCE.	WHERE FOUND.
Oxygen	Lungs and blood.
Hydrogen	Stomach and intestines.
Nitrogen	Blood and intestines.
Carbonic anhydride	Expired air of lungs.
Carburetted hydrogen $\big\}$ Sulphuretted hydrogen	... Lungs and intestines.
Water	Found in all solids and fluids.
Sodium chloride	In all fluids and solids except enamel.
Potassium chloride	In muscles, liver, saliva, gastric juice, etc.
Ammonium chloride	Gastric juice, saliva, tears, urine.
Calcium chloride	Bones, teeth, urine.
Calcium carbonate	Bones, teeth, cartilage, internal ear, blood.
Calcium phosphate $\Big\}$ Magnesium phosphate Sodium phosphate Potassium phosphate	... In all fluids and solids of the body.
Sodium sulphate $\big\}$ Potassium sulphate.	... Universal, except milk, bile and gastric juice.
Sodium carbonate $\big\}$ Potassium carbonate	... Blood, bones, lymph, urine, etc.
Magnesium carbonate	Blood and sebaceous matter.

The inorganic principles enter and leave the body under their own form. *Water* is an essential constituent of all the tissues of the body, constituting about 70 per cent. of the entire body weight. It is introduced into the body in the form of drink and as a constituent of all kinds of food. The average quantity consumed daily is about four pints. While in the body, water acts as a general solvent, gives pliability to various tissues, and promotes the passage of inorganic and organic matters through animal membranes. It also promotes chemical changes which are essential to absorption and assimilation of food and the elimination of products of waste. It is probable that water is also formed within the body by the union of oxygen with the surplus hydrogen of the food. It is eliminated by the skin, lungs and kidneys.

Sodium chloride is present in all the solids and fluids of the body, with the exception of enamel. It regulates osmotic action, holds the albuminous principles of the blood in solution, and preserves the form and consistence of blood corpuscles and the cellular elements of the tissues, by regulating the amount of water entering into their composition.

Calcium phosphate is the most abundant of all the inorganic principles with the exception of water, and is present to a great extent in bone, teeth, muscles and milk. It gives the requisite consistency and solidity to the different tissues and organs. In the blood, it is held in solution by the albuminous constituents.

The *Sodium* and *Potassium phosphates* are present in most of the solids and fluids, and give to them their alkaline reaction. They are chiefly derived from the food.

II. ORGANIC NON-NITROGENIZED PRINCIPLES.

The organic non-nitrogenized principles are derived mainly from the vegetable world, but are also produced within the animal body. They are divided into: 1st, the *carbo-hydrates*, comprising starch and sugar, bodies in which the oxygen and hydrogen exist in the proportion to form water, the amount of carbon being variable; 2d, the *hydro-carbons*, comprising fats, bodies having the same elements entering into their composition, but with the carbon and hydrogen increased and the oxygen diminished in amount.

SUGARS. C. O. H.

Glycogen, or Liver sugar.
Lactose,　or Milk sugar.
Glucose,　or Grape sugar.
Inosite,　or Muscle sugar.

Sugar is found in many of the tissues and fluids of the body; *e.g.*, liver, milk, placenta, blood, muscles, etc. The varieties of sugar are soluble in water, assume the crystalline form upon evaporation, and are converted into alcohol and carbonic acid by fermentation. Sugar is derived from the food, converted, in the alimentary canal, into glucose, absorbed by the veins of the portal system, and then stored up in the liver, under the form of glycogen. When the system requires sugar, it is again returned to the circulation, and plays its part in the nutritive processes of the body. It is finally oxidized, and thus contributes to the formation of heat. It is finally eliminated under the form of carbonic acid and water. There is no experimental proof that sugar contributes directly to the formation of fat in the animal body.

FATS. C. O. H.

Palmitin,　⎫　　　　　　　　Palmitic acid,　⎫
Stearin,　⎬ Neutral Fats.　　Stearic acid,　⎬ Fatty Acids.
Olein,　　⎭　　　　　　　　Oleic acid,　　⎭

The *Neutral fats*, when combined in proper proportions, constitute a large part of the fatty tissue of the body; they are soluble in ether, chloroform and hot alcohol; insoluble in cold alcohol and water, and liquefy at a high temperature; when a neutral fat is subjected to a high temperature in the presence of water and an alkali, it is decomposed, with the assimilation of the elements of water, into a fatty acid and glycerine. The fatty acid combines with the alkali and forms an oleate, palmitate or stearate, according to the fat used. A similar decomposition of the neutral fats is said to take place in the small intestine during digestion. When thoroughly mixed with pancreatic juice, the fats are reduced to a condition of *emulsion*, a state in which the fat is minutely subdivided and the small globules held in suspension.

The *Fatty acids* combined with sodium, potassium and calcium, are found as salts in various fluids of the body, such as blood, chyle, fæces, etc. Phosphorized fats in nervous tissue, butyric acid in milk, propionic acid in sweat, are also constituents of the body.

The Fats are derived from the food, both animal and vegetable. They are deposited in the form of small globules in the cells of the different tissues, are suspended in various fluids, are deposited in masses in and around various anatomical structures and beneath the skin. Independent of the fat consumed as food, there is good experimental evidence that fat is also produced within the animal body from a partial decomposition of the albuminous compounds. Fat serves as a non-conductor of heat, gives roundness and form to the body, and protects various structures from injury. The fats are ultimately oxidized, thus giving rise to heat and force, and are finally eliminated as carbonic acid and water.

III. ORGANIC NITROGENIZED PRINCIPLES.

ALBUMENS. C. O. H. N. S. P.

Albumen.	Myosin.	Mucin.
Albuminose.	Protagon.	Chondrin.
Fibrin.	Pepsin.	Elastin.
Casein.	Pancreatin.	Keratin.
Ostein.	Salivin.	Globulin.

The *Albuminous compounds* are organic in their origin, being derived from the animal and vegetable world; they are taken into the body as food, appropriated by the tissues, and constitute their organic basis; they differ from the non-nitrogenized substances in not being crystalline, but amorphous, in having a more complex but just as definite composition, and containing in addition to C. O. H., nitrogen, with, at times, sulphur and phosphorus. The albumens possess characteristics which distinguish them from all other substances: viz., a *molecular mobility*, which permits *isomeric* modifications to take place with great facility; a *catalytic influence*, in virtue of which they promote, under favorable conditions, chemical changes in other substances: *e.g.*, during digestion, salivin and pepsin cause starch an albumen to be transformed into sugar and albuminose respectively. Different albumens possess varying proportions of water, which they lose when subjected to desiccation, becoming solid; but upon exposure to moisture they again absorb water, regaining their original condition—they are *hygroscopic*. Another property is that of *coagulation*, which takes place under certain conditions: *e. g.*, the presence of mineral acids, heat, alcohol, etc.

After death the albuminous compounds undergo putrefactive changes, giving rise to carburetted and sulphuretted hydrogen and other gases.

Albumen exists in the blood, lymph, chyle, constituting the pabulum of the tissues; it is coagulated by heat, mineral acids and alcohol.

Peptones are found in the stomach from the digestion of albuminous principles of the food; they are coagulated by tannic acid, chlorine, acetate of lead, and characterized by great diffusibility, which permits them to pass through animal membranes with facility.

Fibrin can be obtained from freshly drawn blood by whipping; it also coagulates spontaneously, and when examined microscopically exhibits a filamentous structure.

Casein is the albuminous principle of milk.

Ostein constitutes the organic basis of bone, with which are mingled the **salts of lime.**

Myosin is found in muscles, **protagon** in brain, *pepsin, pancreatin and* **salivin,** in the digestive fluids.

Mucin, *chondrin, elastin, keratin and globulin*, are found in mucus, **cartilage,** elastic tissues, hair, nails, and red corpuscles, respectively.

As the properties of the compounds formed by the union of elements are the resultants of the properties of the elements themselves, it follows that the ternary substances, sugars, starches and fats, possess a great inertia and a notable instability; while in the more complex albuminous compounds, in which sulphur and phosphorus are united to the four chief elements, molecular mobility, resulting in isomerism, exists in a high degree. As these compounds are unstable, of a great molecular mobility, they are well fitted to take part in the composition of organic bodies, in which there is a continual movement of composition and decomposition.

IV. PRINCIPLES OF WASTE.

Urea,	Xanthin,	Sodium,	
Creatin,	Tyrosin,	Potassium,	Urates.
Creatinin,	Hippuric Acid,	Ammonium,	
Cholesterin,	Calcium Oxalate,	Calcium,	

These principles, which represent waste, are of organic origin, arising within the body as products of disassimilation or retrograde metamorphosis of the tissues; they are absorbed by the blood, carried to the various excretory organs, and by them eliminated from the body.

The excrementitious substances will be fully considered under excretion.

Proximate Quantity of the Chemical Elements **and Proximate** Principles of the Body, Weighing 154 **lbs.**

	lbs.	oz.		lbs.	oz.
Oxygen	111	...	Water	111	...
Hydrogen	14	...	Albuminoids	23	7
Nitrogen	3	8	Fats	12	...
Carbon	21	...	Calcium phosphate	5	13
Calcium	2	...	Calcium carbonate	1	..
Phosphorus	1	12	Calcium fluoride	...	3
Sodium, etc	...	12	Sodium sulphate, etc	...	9
	154	...		154	...

STRUCTURAL COMPOSITION of THE BODY.

The Study of the Structure of the body reveals that it is composed of dissimilar parts, *e. g.*, bones, muscles, nerves, lungs, etc.; while these, again, by closer examination, can be resolved into elementary structures, the *tissues, e.g.*, connective tissue, muscular, nervous, epithelial tissue, etc.

Microscopical examination of the tissue shows that they are composed of fundamental structural elements, termed *cells*.

Cells are living physiological units; **the simplest structural forms capable** of manifesting the phenomena of life.

Cells vary in their anatomical constitution in the different structures of the body, and may be classed in three groups, viz.: 1. Cells possessing a distinct *cell wall, cell substance* and a *nucleus*. 2. Cells possessing a *cell substance* and a *nucleus*. 3. Cells possessing the *cell substance* only. They vary in size, from the $\frac{1}{3600}$ to the $\frac{1}{300}$ of an inch in diameter; when young and free to move in a fluid medium they assume the spherical form; but when subjected to pressure, may become flattened, cylindrical, fusiform or stellate.

Structure of Cells. The cell wall is not an essential structure, as many cells are entirely devoid of it. It is a thin, structureless, transparent membrane, permeable to fluids.

The Cell Substance in young cells is a soft, viscid, albuminous matter, unstable, insoluble in water, and known as *protoplasm, bioplasm, sarcode,* etc.; in older cells the original cell substance undergoes various transformations, and is partly replaced by fat globules, pigment and crystals.

The Nucleus is a small vesicular body in the interior of the cell substance, and frequently contains smaller bodies, the *nucleoli.*

MANIFESTATIONS OF CELL LIFE.

Growth. Cells when newly formed are exceedingly small, but as they approach maturity they increase in size, by the capability which the cells possess of selecting and appropriating new material as food, vitalizing and organizing it. The extent of cell growth varies in different tissues; in some the cells remain exceedingly small, in others they attain considerable size. In many instances the cell substance undergoes transformation into new compounds destined for some ulterior purpose.

Reproduction. Like all organic structures cells have a limited period of life; their continual decay and death necessitates a capability of reproduction. Cells reproduce themselves in the higher animals mainly by *fission.* This is seen in the white blood corpuscles of the young embryos of animals; the corpuscle here consists of a cell substance and nucleus. When division of the cell is about to take place, the nucleus elongates, the cell substance assumes the oval form, a constriction occurs, which gradually deepens, until the original cell is completely divided and two new cells are formed, each of which soon grows to the size of the parent cell.

In cells provided with a cell membrane the process is somewhat different. In the ova of the inferior animals, after fertilization has taken place, a furrow appears on the opposite sides of the cell substance, which deepens until the cell is divided into two equal halves, each containing a nucleus; this process is again repeated until there are four cells, then eight, and so on until the entire cell substance is divided into a mulberry mass of cells, completely occupying the interior of the cell membrane. The whole process of segmentation takes place with great rapidity, occupying not more than a few minutes, in all probability.

Motion. Spontaneous movement has been observed in many of the cells of the body. It may be studied, for example, in the movements of the spermatozoids, the waving of the cilia covering the cells of the bronchial mucous membrane, the white corpuscles of the blood, etc.

By a combination and transformation of these original structural elements, and material derived from them, all the tissues are formed which enter into the structure of the human body.

CLASSIFICATION OF TISSUES.

I. Homogeneous Substance, a more or less solid, albuminous structure, filling the spaces between the cells and fibres of various tissues, *e. g.*, cartilage, bone, dentine, etc.

II. Limiting Membrane, a thin, homogeneous membrane, structureless, composed of coagulated albumen, and often not more than the $\frac{1}{20000}$ of an inch in thickness, found lining the blood vessels and lymphatics, forming the basement membrane of the skin and mucous membranes, the posterior layer of the cornea, the capsule of the crystalline lens, etc.

III. Simple fibrous or **filamentous tissue**—the elements of which **are real** or apparent filaments.

(a) *Connective or areolar; white fibrous tissue;* constituting tendons, ligaments, aponeuroses, periosteum, dura mater, synovial membranes, vascular tunics, etc.

(b) *Yellow elastic tissue,* found in the middle coats of arteries, **veins,** lymphatics, ligamentum nuchæ, vocal cords, ligamenta subflava, etc.

IV. Compound membranes (membrano-cellular or **fibro-cellular tissues**), cells aggregated into laminæ.

(a) *Epidermic tissue;* (b) *epithelial tissue;* (c) *glandular tissue;* (d) *cornea.*

V. Cells containing coloring matter, or pigment cells, *e. g.*, skin, choroid membrane, etc.

VI. Cells coalesced or consolidated by internal deposits, *e. g.*, hair, nails, bone, teeth, etc.

VII. Cells imbedded in an intercellular substance, *e. g.*, cartilage, crystalline lens, etc.

VIII. Cells aggregated in clusters, forming tissues more or less solid, *e. g.*, adipose tissue, lymphatic glands.

IX. Cells imbedded in a matrix of capillaries, *e. g.*, gray or vesicular nervous matter.

X. Cells whose coalesced cavities form tubes containing liquids or secondary solid deposits, *e. g.*, vascular tissue, dentine.

XI. Cells free, isolated, or floating—fluid tissue—*e. g.*, red and white blood corpuscles, lymph and chyle corpuscles.

FOOD.

A **Food** may be defined to be any substance capable of playing a part in the nutrition of the body.

Food is required for the repair of the waste of the tissues consequent on their functional activity, for the generation of heat and the evolution **of force.**

Hunger and Thirst are sensations which indicate the necessity for taking food; they arise in the tissues at large, and are referred to the stomach and fauces, respectively, through the sympathetic nervous system.

Inanition or Starvation results from an insufficiency or absence of food, the physiological effects of which are hunger, intense thirst, intestinal uneasiness, weakness and emaciation; the quantity of carbonic acid exhaled diminishes and the urine is lessened in amount; the volume of the blood diminishes; a fetid odor is exhaled from the body; vertigo, stupor followed by delirium, and at times convulsions, result from a disturbance of the nerve centres; a marked fall of the bodily temperature occurs, from a diminished activity of the nutritive process. Death usually takes place, from exhaustion.

During starvation the **loss of different tissues, before death occurs, aver-**ages $\frac{4}{10}$, or 40 per cent. **of their weight.**

Those tissues which lose *more* than *40 per cent.* **are fat,** 93.3; blood, 75; spleen, 71.4; pancreas, 64.1; liver, 52; heart, 44.8; intestines, 42.4; muscles, 42.3. Those which **lose** *less* than *40 per cent.* are the muscular coat of the stomach, 39.7; pharynx and œsophagus, 34.2; skin, 33.3; kidneys, **31.9**; respiratory apparatus, 22.2; bones, 16.7; eyes, 10; nervous system, 1.9.

The *Fat* entirely disappears, with the exception of a small quantity which remains in the posterior portion of the orbits and around the kidneys. The *Blood* diminishes in volume and loses its nutritive properties. The *Muscles* undergo a marked diminution in volume and become soft and flabby. The *Nervous system* is last to suffer, not more than two per cent. disappearing before death occurs.

The *appearances* presented by the body after death from starvation **are those** of anæmia and great emaciation; almost total absence of fat; bloodlessness; a diminution in the volume of the organs; an empty condition of the stomach and bowels, the coats of which are thin and transparent. There is a marked disposition **of the** body to undergo decomposition, giving rise to a very fetid odor.

The *duration of life* after a complete deprivation of food **varies** from eight to thirteen days, though life can be maintained much longer if a quantity of water be obtained. The water is more essential under these circumstances than the solid matters, which can be supplied by the organism itself.

The food consumed daily is a heterogeneous compound consisting of both nutritious and innutritious portions. The nutritious portions are known as the alimentary principles, while the food, as a whole, is known as aliment.

The different *alimentary principles* which are appropriated by the system are combined in different proportions in the various articles of food, and are separated from the innutritious substances during the process of digestion. They belong to the organic and inorganic worlds, and may be classified, according to their chemical composition, as follows :—

CLASSIFICATION OF ALIMENTARY PRINCIPLES.

1. Albuminous group—nitrogenized, C. O. H. N. S. P.

PRINCIPLE.	WHERE FOUND.
Myosin, syntonin	Flesh of animals.
Vitellin, albumen	Yolk of egg, white of egg.
Fibrin, globulin	Blood contained in meat.
Casein	Milk, cheese.
Gluten	Grain of wheat and other cereals.
Vegetable albumen	Soft growing vegetables.
Legumin	Peas, beans, lentils, etc.
Gelatin	Bones.

2. Saccharine group—non-nitrogenized, C. O. H.

Cane sugar, beet root sugar	Sugar cane, beets, etc.
Glucose, grape sugar	Fruits.
Inosite, liver sugar, glycogen	Muscles, liver, etc.
Lactose or milk sugar	Milk.
Starch	Cereals, tuberous roots and leguminous plants.

3. Oleaginous group—non-nitrogenized, C. O. H.

Animal fats and oils ⎫	Found in the adipose tissue of animals,
Stearin, olein ⎬	seeds, grains, nuts, fruits, and other
Palmatin, fatty acids ⎭	vegetable tissues.

4. Inorganic group. Water, sodium and potassium chlorides, sodium, calcium, magnesium and potassium phosphates, calcium carbonate and iron.

5. Vegetable acid group. Malic, citric, tartaric and other acids, found principally in fruits.

6. Accessory foods. Tea, coffee, alcohol, cocoa, etc.

The *Albuminous principles* enter largely into the composition of the body, and constitute the organic bases of the different tissues; they are mainly required for the growth and **repair** of the tissues. There is good reason to believe that the albuminous **principles** are decomposed in the **body into** fat and urea, and **the former when oxidized** gives rise to the evolution of heat and force, while the latter is eliminated **by the** kidneys. Muscular work, however, does not result from a destruction of **the** albuminous compounds. The oxidation of the carbonaceous compounds, sugars and oils, furnishing the force which is transformed by the muscular system into motor power. When employed exclusively **as** food for any length of time, the albuminous substances are incapable of supporting life.

The *Saccharine principles* are important to the process of nutrition, but **the** changes **which** they undergo are **not** fully understood; they form but a small proportion of the animal tissues, and **by** oxidation generate heat **and** force. *Starch* undergoes conversion **into dextrin** and grape sugar.

The *Oleaginous principles* form a large part of the tissues of the **body.** They are introduced into the system as food, and are formed also from **a** transformation of albuminous matter during the nutritive process; they enter into the composition of nervous and muscular tissue, and are stored up as adipose tissue in the visceral cavities and subcutaneous connective tissue, thus giving roundness to the form and preventing, to some extent, the radiation of heat. While they aid **in the** reconstruction of tissue, they **mainly** undergo oxidation, giving rise **to the** production of heat and the **evolution of** muscular **and** nervous force.

The *Inorganic principles* constitute an essential part of all animal tissues, **and are introduced with the food.**

Water **is** present in all fluids and solids of the body, holding their ingredients **in** solution, promoting the absorption of new material into the blood and tissues, and the removal of waste ingredients.

Sodium chloride is an essential constituent of all tissues, regulating the passage of fluids through animal membranes (endosmosis and exosmosis).

Calcium phosphate gives solidity to bones and teeth, constituting more than one-half their substance.

Iron **is** a constituent of the coloring matter of the blood.

The *Vegetable acids* are important to nutrition, and tend to prevent the scorbutic diathesis.

The *Accessory foods* also influence the process of nutrition. *Tea* excites the respiratory function, increasing the elimination of carbonic acid. *Coffee* is a stimulant to the nervous system; increases the force of the heart's action, increases the arterial tension and retards waste.

Alcohol, when introduced into the system in *small quantities*, undergoes oxidation and contributes to the production of force, and is thus far a food. It excites the gastric glands to increased secretion, improves the digestion, accelerates the action of the heart and stimulates the activities of the nervous centres. In zymotic diseases, and all cases of depression of the vital powers, it is most useful as a restorative agent. When taken in *excessive quantities*, it is eliminated by the lungs and kidneys. The metamorphosis of the tissues is retarded, the elimination of urea and carbonic acid is lessened, the temperature lowered, the muscular powers impaired and the resistance to depressing external influences diminished. When taken through a long period of time, alcohol impairs digestion, produces gastric catarrh, disorders the secreting power of the hepatic cells. It also diminishes the muscular power and destroys the structure and composition of the cells of the brain and spinal cord. The connective tissue of the body increases in amount, and subsequently contracting, gives rise to sclerosis.

A proper combination of different alimentary principles is essential for healthy nutrition; no one class being capable of maintaining life for any definite length of time.

The *Albuminous food* in excess promotes the arthritic diathesis, manifesting itself as gout, gravel, etc.

The *Oleaginous food* in excess gives rise to the bilious diathesis, while a deficiency of it promotes the scrofulous.

The *Farinaceous food*, when long continued in excess, favors the rheumatic diathesis by the development of lactic acid.

The Alimentary Principles are not introduced into the body as such, but are combined in proper proportions to form compound substances, termed *foods*, *e.g.*, bread, milk, eggs, meat, etc., the nutritive value of each depending upon the extent to which these principles exist.

PERCENTAGE COMPOSITION OF DIFFERENT FOODS.

	WATER.	ALBUMEN.	STARCH.	SUGAR.	FATS.	SALTS.
Bread	37	8.1	47.4	3.6	1.6	2.3
Milk	86	4.1	...	5.2	3.9	0.8
Eggs,	74	14.0	...	...	10.5	1.5
Meat	54	27.6	...	...	15.45	2.95
Potatoes	75	2.1	18.8	3.2	0.2	0.7
Corn,	14	11.1	64.7	0.4	8.1	1.7
Oatmeal	15	12.6	58.4	5.4	5.6	3
Turnips	91	1.2	5.1	2.1	...	6
Carrots	83	1.3	8.4	6.1	0.2	1.0
Rice	13	6.3	79.1	0.4	0.7	0.5

The amount of food required in 24 hours is estimated from the total quantity of carbon and nitrogen excreted from the body in 24 hours; these two elements representing the waste or destruction of the carbonaceous and nitrogenized compounds. It has been determined by experimentation that about 4600 grains of carbon and about 300 grains of nitrogen are eliminated from the body daily; the ratio being about 15 to 1. That the body may be kept in its normal condition, a proper proportion of carbonaceous (bread) to nitrogenized (meat) food should be observed in the diet.

The method of determining the proper amounts of both kinds of food is as follows:—

1000 grains of bread (2 oz.) contain 300 grs. C. and 10 grs. N.

To obtain the requisite amount of nitrogen from bread, 30,000 grains, or about 4 lbs., containing 9000 grains of carbon and 300 of nitrogen, would have to be consumed. Under such a diet there would be a large excess of carbon, which would be undesirable. On a meat diet the reverse obtains:—

1000 grains of meat (2 oz.) contain 100 grs. C. and 30 grs. N.

To obtain the requisite amounts of carbon from meat, 45,000 grains, or about 6½ lbs., containing 4500 grains of carbon and 1350 grains of nitrogen, would have to be consumed. Under such circumstances there would arise an excess of nitrogen in the system, which would be equally undesirable and injurious. By combining these two articles, however, in proper proportion, the requisite amounts of carbon and nitrogen can be obtained without any excess of either, e. g. :—

$$
\begin{array}{llll}
\text{2 lbs. of bread contain} & \text{4630 grs. C. and} & \text{154 grs. N.} \\
\text{¾ “ \quad meat \quad “} & \text{463 “ “ “} & \text{154 “ “} \\
\hline
& \text{5093 C.} & \text{308 N.}
\end{array}
$$

The amount of carbon and nitrogen necessary to compensate for the loss to the system daily would be contained in the above amounts of food. As about 3½ oz. of oil or butter are consumed daily, the quantity of bread can be reduced to 19 oz. In the quantities of bread and meat above mentioned, there are 4.2 oz. albumen, 9.3 oz. sugar and starch.

COMPARISON OF INGESTA AND EGESTA IN 24 HOURS.

FOOD, DRINK, AIR.	OZ.	EXCRETIONS.	OZ.
Albumen	4.23	Breath { carbonic acid } { watery vapor }	43.40
Starch	11.63		
Fats	3.17	Perspiration { carbonic acid } { watery vapor }	23.62
Salts	1.13		
Water (6 pints)	93.00	Urine	66.31
Oxygen	26.24	Solid excreta	6.07
Total	139.40	Total	139.40

DIGESTION.

Digestion is a physical and chemical process, by which the food introduced into the alimentary canal is liquefied and its nutritive principles transformed by the digestive fluids into new substances capable of being absorbed into the blood.

The Digestive Apparatus consists of the alimentary canal and its appendages, viz.: teeth, salivary, gastric and intestinal glands, liver and pancreas.

Digestion may be divided into *seven stages :* prehension, mastication, insalivation, deglutition, gastric and intestinal digestion and defecation.

Prehension, the act of conveying food into the mouth, is accomplished by the hands, lips and teeth.

Mastication is the trituration of the food, and is accomplished by the teeth and lower jaw, under the influence of muscular contraction. When thoroughly divided, the food presents a greater surface for the solvent action of the digestive fluids, thus aiding the general process of digestion.

The Teeth are thirty-two in number, sixteen in each jaw, and divided into four incisors or cutting teeth, two canines, four bicuspids, and six molars or grinding teeth; each tooth consists of a crown covered by enamel, a neck, and a root surrounded by the crusta petrosa, and imbedded in the alveolar process; a section through a tooth shows that its substance is made of *dentine*, in the centre of which is the pulp cavity, containing blood vessels and nerves.

The *lower jaw* is capable of making a downward and an upward, a lateral and an antero-posterior movement, dependent upon the construction of the temporo-maxillary articulation.

The jaw is *depressed* by the contraction of the *digastric, genio-hyoia, mylo-hyoid* and *platysma myoides* muscles; *elevated* by the *temporal*, *masseter* and *internal pterygoid* muscles; moved *laterally* by the alternate contraction of the *external pterygoid* muscles; moved *anteriorly* by the *pterygoid* and *posteriorly* by the united actions of the *genio-hyoid, mylo-hyoid* and posterior fibres of the *temporal* muscle.

The food is kept between the teeth by the *intrinsic* and *extrinsic* muscles of the tongue from within, and the *orbicularis oris* and *buccinator* muscles from without.

The **Movements of Mastication**, though originating in an effort of the will and under its control, are, for the most part, of an automatic or reflex character, taking place through the medulla oblongata and induced by the presence of food within the mouth. The nerves and nerve centres involved in this mechanism are shown in the following table :—

NERVOUS CIRCLE OF MASTICATION.

AFFERENT OR EXCITOR NERVES.	EFFERENT OR MOTOR NERVES.
1. Lingual branch of 5th pair.	1. 3d branch of 5th pair.
2. Glosso-pharyngeal.	2. Hypo-glossal.
	3. Facial.

The impressions made upon the terminal filaments of the sensory nerves are transmitted to the medulla; motor impulses are here generated which are transmitted through motor nerves to the muscles involved in the movements of the lower jaw. The medulla not only generates motor impulses, but coördinates them in such a manner that the movements of mastication may be directed toward the accomplishment of a definite purpose.

Insalivation is the incorporation of the food with the saliva secreted by the *parotid, sub-maxillary* and *sub-lingual glands;* the *parotid* saliva, thin and watery, is poured into the mouth through Steno's duct; the *sub-maxillary* and *sub-lingual* salivas, thick and viscid, are poured into the mouth through Wharton's and Bartholini's ducts.

In their minute structure the salivary glands resemble each other. They belong to the racemose variety, and consist of small sacs or vesicles, which are the terminal expansions of the smallest salivary ducts. Each vesicle or *acinus* consists of a basement membrane surrounded by blood vessels and lined with epithelial cells. In the parotid gland the lining cells are granular and nucleated; in the sub-maxillary and sub-lingual glands the cells are large, clear, and contain a quantity of mucigen. During and after secretion very remarkable changes take place in the cells lining the acini, which are in some way connected with the essential constituents of the salivary fluids.

In a living serous gland, *e. g.*, parotid, during rest, the secretory cells lining the acini of the gland are seen to be filled with fine granules, which are often so abundant as to obscure the nucleus and enlarge the cells until the lumen of the acinus is almost obliterated (Fig. 1). When the gland begins to secrete the saliva, the granules disappear from the outer boundary

FIG. 1.

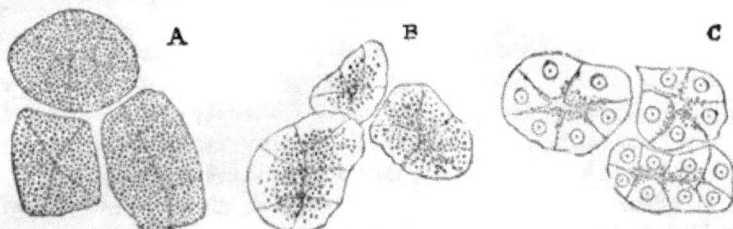

Cells of the alveoli of a serous or watery salivary gland. A. After rest. B. After a short period of activity. C. After prolonged period of activity.— *From Yeo's Text-Book of Physiology.*

of the cells, which then become clear and distinct. At the end of the secretory activity, the cells have become free of granules, have become smaller and more distinct in outline. It would seem that the granular

FIG. 2.

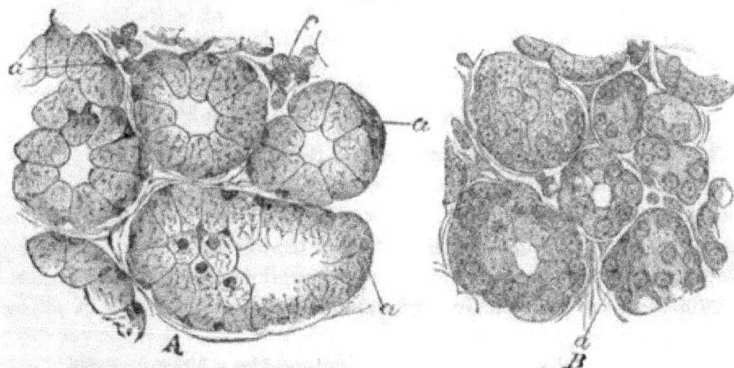

Section of a " mucous " gland. A. In a state of rest. B. After it has been for some time actively secreting.—*After Lavdowsky.*

matter is formed in the cells during the rest, and discharged into the ducts during the activity of the gland.

In the mucous glands, *e. g.*, sub-maxillary and sub-lingual, the changes that occur in the cells are somewhat different (Fig. 2). During the inter-

C

vals of digestion, the cells lining the gland are large, clear and highly refractive, and contain a large quantity of mucigen. After secretion has taken place, the cells exhibit a marked change. The mucigen cells have disappeared, and in their place are cells which are small, dark and composed of protoplasm. It would appear that the cells, during rest, elaborate the mucigen which is discharged into the tubules during secretory activity, to become part of the secretion.

Saliva is an opalescent, slightly viscid, alkaline fluid, having a specific gravity of 1.005. Microscopical examination reveals the presence of salivary corpuscles and epithelial cells. Chemically it is composed of water, proteid matter, a ferment (*ptyalin*) and inorganic salts. The amount secreted in 24 hours is about 2½ lbs. Its function is twofold :—

1. *Physical.*—Softens and moistens the food, glues it together, and facilitates swallowing.

2. *Chemical.*—Converts *starch* into *grape sugar*. This action is due to the presence of the organic ferment, ptyalin. The change consists in the assumption of a molecule of water.

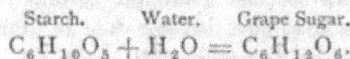

$$\underset{\text{Starch.}}{C_6H_{10}O_5} + \underset{\text{Water.}}{H_2O} = \underset{\text{Grape Sugar.}}{C_6H_{12}O_6}.$$

NERVOUS CIRCLE OF INSALIVATION.

AFFERENT OR EXCITOR NERVES.

1. Lingual branch of 5th pair.
2. Glosso-pharyngeal.

EFFERENT OR MOTOR NERVES.

1. Auriculo-temporal branch of 5th pair, for parotid gland.
2. Chorda tympani, for sub-maxillary and sub-lingual glands.

The centres regulating the secretion are two, viz. : The medulla oblongata and the sub-maxillary ganglion of the sympathetic ; the latter acting antagonistically to the former. Impressions excited by the food in the mouth reach the medulla oblongata through the afferent nerves ; motor impulses are there generated which pass outward through the efferent nerves.

Stimulation of the auriculo-temporal branch increases the flow of saliva from the parotid gland ; *division* arrests it.

Stimulation of the chorda tympani is followed by a dilation of the blood vessels of the sub-maxillary gland, increased flow of blood (thus acting as a vaso-dilator nerve) and an abundant discharge of a thin saliva; *division* of the nerve arrests the secretion.

Stimulation of the cervical sympathetic is followed by a contraction of the blood vessels, diminishing the flow of blood (thus acting as a vaso-constrictor nerve) and a diminution of the secretion, which now becomes thick and viscid; *division* of the sympathetic does not, however, completely dilate the vessels. There is evidence of the existence of a local vaso-motor mechanism, which is *inhibited* by the chorda tympani; *exalted* by the sympathetic.

Deglutition is the act of transferring food from the mouth into the stomach, and may be divided into *three* stages :—

1. The passage of the bolus from the mouth into the pharynx.

2. From the pharynx into the œsophagus.

3. From the œsophagus into the stomach.

In the *1st stage*, which is entirely voluntary, the mouth is closed and respiration momentarily suspended; the tongue, placed against the roof of the mouth, arches upward and backward, and forces the bolus into the fauces.

In the *2d stage*, which is entirely reflex, the palate is made tense and directed upward and backward by the levatores-palati and tensores-palati muscles; the bolus is grasped by the superior constrictor muscle of the pharynx and rapidly forced into the œsophagus.

The food is prevented from entering the *posterior nares* by the uvula and the closure of the posterior half-arches (the palato-pharyngei muscles); from entering the *larynx* by its ascent under the base of the tongue and the action of the epiglottis.

In the *3d stage*, the longitudinal and circular muscular fibres, contracting from above downward, *strip* the bolus into the stomach. [For nervous mechanism of Deglutition, see **Medulla** Oblongata.]

Gastric Digestion. The stomach is a dilation of the alimentary canal, 13 inches long, 5 inches deep, having a capacity of about 5 pints; there can be distinguished a cardiac and pyloric orifice, a greater and lesser curva-ture, a greater and lesser pouch.

It possesses three coats :—

1. Serous, a reflection of the peritoneum.

2. Muscular, the fibres of which are arranged longitudinally, transversely and obliquely.

3. Mucous, thrown into folds, forming the rugæ.

Imbedded in the mucous coat are immense numbers of *mucous* and *true gastric glands*. In the pyloric end of the stomach are found the *mucous* glands, which are lined with columnar epithelium throughout their extent. In the cardiac end are found the true *peptic glands* (Fig. 3), the ducts of which are also lined with columnar cells, while the secretory parts are lined with two distinct varieties of cells. One variety consists of small spheroidal, granular cells, which border the lumen of the gland, and are known as the *chief cells;* the other **variety consists of** large, oval, well-defined granular cells, much less abundant, and **are situated** between the basement membrane of the **gland and the** chief cells. **From their** position they have been termed *parietal cells.* During the intervals of digestion **the** chief cells are pale, and the hyaline substance of which they are com-**posed is finely granular.** During the **stage** of active secretion the **cells**

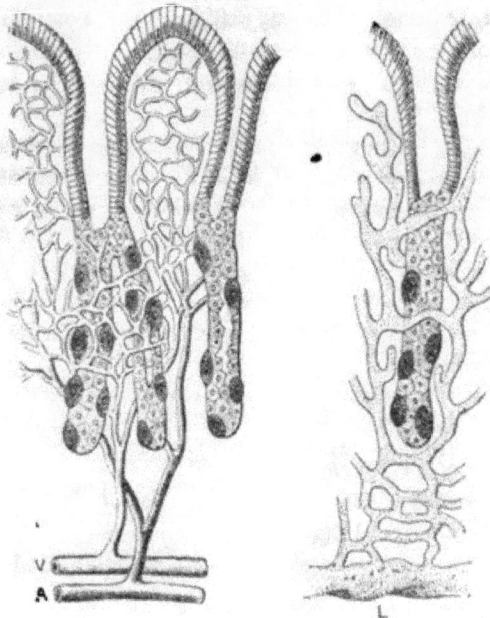

FIG. 3.

Diagram showing the relation of the ultimate twigs of the blood vessels, V and A, and of the absorbent radicles to the glands of the stomach and the different kinds of epithelium, viz., above cylindrical cells; small, pale cells in the lumen, outside which are the dark ovoid cells.—*From Yeo's Text-Book of Physiology.*

become swollen and turbid, and are then said to be rich in pepsin. Toward the end of digestion the granules disappear, the cells become pale and return to their former size.

During the intervals of digestion, the mucous membrane of the stomach is pale and covered with a layer of mucus. Upon the introduction of food, the blood vessels dilate and become filled with blood, and the mucous membrane becomes red. At the same time small drops of a fluid, the gastric juice, begin to exude upon its surface, which gradually run together and trickle down the sides of the stomach.

The secretion of gastric juice is a reflex act, taking place through the central nervous system and called forth in response to the stimulus of food in the stomach. That the central nervous system also directly influences the production of the secretion is shown by the fact that mental emotion, such as fear and anger, will arrest or vitiate the normal secretion. The reflex nature of the process can be shown by experimentation upon the pneumo-gastric nerve. If during digestion, when the peristaltic movements are active and the gastric mucous membrane flushed and covered with gastric juice, the pneumogastric nerves are divided on both sides, the mucous membrane becomes pale, the secretion is arrested and the peristaltic movements become less marked. Stimulation of the peripheral end produces no constant effects ; stimulation of the central end, however, is at once followed by dilatation of the vessels, flushing of the mucous membrane and a re-establishment of the secretion. It is evident, therefore, that during digestion afferent impulses are passing up the pneumogastrics to the medulla ; efferent impulses, in all proba-bility, pass through the fibres of the sympathetic nervous system to the blood vessels and glands concerned in the elaboration of the gastric juice. After all the nervous connections of the stomach are divided, a small quantity of juice continues to be secreted for several days. This has been attributed to the action of a local nervous mechanism and to the direct action of the food upon the protoplasm of the secreting cells.

The Gastric Juice is a secretion of the true peptic glands, and when obtained from the stomach through a fistulous opening, is a clear, straw-colored fluid, decidedly acid, with a specific gravity of 1.005 to 1.010.

COMPOSITION OF GASTRIC JUICE.

Water	975.00
Pepsin	15.00
Hydrochloric acid	4.78
Inorganic salts	5.22
	1000.00

The water forms the largest part of the fluid, and holds in solution the other ingredients. It results from a transudation from the blood vessels under the increased blood supply. Of the inorganic salts the chlorides of sodium and potassium are the most abundant.

Pepsin is the organic nitrogenized ferment of the gastric juice, and is formed, during the intervals of digestion, by the peptic cells. In the presence of a small per cent. of an acid, it acquires the property of converting the albumen of the food into albuminose or peptones.

Hydrochloric acid is present in small quantity, and gives the juice its acidity. In all probability, its production is due to the activity of the parietal cells. These two characteristic ingredients of the gastric juice exist in a state of combination as hydrochloro-peptic acid, and the presence of both is absolutely essential for the complete digestion of the food.

When the food enters the stomach, it is subjected to the peristaltic action of the muscular coat, and thoroughly incorporated with the gastric juice. This fluid has a twofold action upon the food:—1st. A physical action, by which the fibrous tissues of meats, the cellulose and hard parts of grains and vegetables, are dissolved away until the food is disintegrated and reduced to the liquid condition. 2d. A chemical action, by which the albuminous principles are transformed into peptones. The more important foods with their contained albuminous principles are shown on page 19.

Upon *meat* the gastric juice has a decidedly disintegrating action. The connective tissue is first dissolved, the fibres are separated, the sarcolemma softened, and the whole reduced to a grumous, pultaceous mass. *Milk* undergoes coagulation in from ten to fifteen minutes, the casein being precipitated in the form of soft flocculi, which are easy of transformation into peptone. Upon *Vegetable tissues*, the gastric juice exerts also a disintegrating action; the cellulose and woody fibres are dissolved and the nutritive principles liberated. *Bread* undergoes liquefaction quite readily.

The **Principal** Action of the gastric juice, however, is to transform the different albuminous principles of the food into *peptones* or *albuminose*, the different stages of which are due to the acid and pepsin respectively. When freed from its combination, the hydrochloric acid converts the albumen into *acid albumen* or *parapeptone;* while this intermediate product is being formed, the pepsin converts it at once into *peptone.* In order that the digestion of albumen may be complete, it is necessary that both the acid and pepsin be present in proper quantity. Before digestion, the albuminous principles are insoluble in water and incapable of being absorbed. After digestion, they become soluble and are readily absorbed. Peptones differ from the albumins in being—

1. *Diffusible*, passing rapidly through the mucous membrane and walls of the blood vessels.

2. *Non-coagulable* by heat, nitric or acetic acids; but are readily precipitated by tannic acid.

3. *Soluble* in water and saline solutions.

4. *Assimilable* by the blood; when injected into it, they do not reappear in the urine.

Gastric juice exerts no influence either upon grape sugar, cane sugar, starch or fat.

Gastric Digestion occupies on the average from 3 to 5 hours, but varies in duration according to the nature and quantity of the food, exercise, temperature, etc.

TABLE SHOWING DIGESTIBILITY OF VARIOUS ARTICLES OF FOOD.

	HOURS.	MINUTES.
Eggs, whipped,	1	30
" soft boiled,	3	
" hard boiled,	3	30
Oysters, raw,	2	55
" stewed,	3	30
Lamb, broiled,	2	30
Veal, broiled,	4	
Pork, roasted,	5	15
Beefsteak, broiled,	3	
Turkey, roasted,	2	25
Chicken, boiled,	4	
" fricasseed,	2	45
Duck, roasted,	4	
Soup, barley, boiled,	1	30
" bean, "	3	
" chicken, "	3	
" mutton, "	3	30
Liver, beef, broiled,	2	
Sausage, "	3	20
Green corn, boiled,	3	45
Beans, "	2	30
Potatoes, roasted,	2	30
" boiled,	3	30
Cabbage, "	4	30
Turnips, "	3	30
Beets "	3	45
Parsnips, "	2	30

The *Amount* of gastric juice secreted in 24 hours varies, under normal conditions, from 8 to 14 pounds.

Movements of the Stomach. As soon as digestion commences, the cardiac and pyloric orifices are closed; the walls of the stomach contract upon the food, and a peristaltic action begins, which carries the food along the greater and lesser curvatures, and thoroughly incorporates it with the

gastric juice. As soon as any portion of the food is digested, it passes
through the pylorus into the intestine.

Vomiting. The act of vomiting is usually preceded by nausea and a
discharge of saliva into the mouth. This is then swallowed, and carries
into the stomach a quantity of air which facilitates the ejection of the contents of the stomach by aiding the relaxation of the cardiac sphincter. A
deep inspiration is then taken, during which the lower ribs are drawn in
and the diaphragm descends and remains contracted. At the same time
the glottis is closed. A sudden expiratory effort is now made, and the
cardiac orifice being open, the abdominal muscles contracting, press upon
the stomach and forcibly eject its contents into the mouth.

Intestinal Digestion. The intestine is about 20 feet long, 1½ inches
in diameter, and possesses three coats :—

    1. Serous (peritoneal).

    2. Muscular, the fibres of which are arranged longitudinally and trans-
versely.

    3. Mucous, thrown into folds, forming the *valvulæ conniventes.*

This stage of digestion is probably the most complex and important;
here the different alimentary principles are further elaborated and prepared
for absorption into the blood by being acted upon by the *intestinal juice,
pancreatic juice* and *bile.*

Throughout the mucous coat are imbedded the intestinal follicles, the
glands of Brunner and Lieberkühn. They secrete the true *intestinal juice,*
which is an alkaline, viscid fluid, composed of water, organic matter and
salts. Its *function* is to convert *starch* into glucose, and assist in the
digestion of the albuminoids.

The Pancreatic Juice is secreted by the pancreas, a flattened gland
about six inches long, running transversely across the posterior wall of the
abdomen, behind the stomach ; its duct opens into the duodenum.

The pancreas is similar in structure to the salivary glands, consisting of
a system of ducts terminating in acini. The acini are tubular or flask-
shaped, and consist of a basement membrane lined by a layer of cylindrical,
conical cells, which encroach upon the lumen of the acini. The cells
exhibit a difference in their structure (Fig. 4), and may be said to consist
of two zones, viz., an *outer parietal zone,* which is transparent and apparently homogeneous, staining rapidly with carmine; an *inner zone,* which
borders the lumen, and is distinctly granular and stains but slightly with
carmine. These cells undergo changes similar to those exhibited by the
cells of the salivary glands during and after active secretion. As soon as

the secretory activity of the pancreas is established, the granules disappear, and the inner granular layer becomes reduced to a very narrow border while the outer zone increases in size and occupies nearly the entire cell. During the intervals of secretion, however, the granular layer reappears and increases in size until the outer zone is reduced to a minimum. It would seem that the granular matter is formed by the nutritive processes occurring in the gland during rest, and is discharged during secretory activity into the ducts and takes part in the formation of the pancreatic secretion.

FIG. 4.

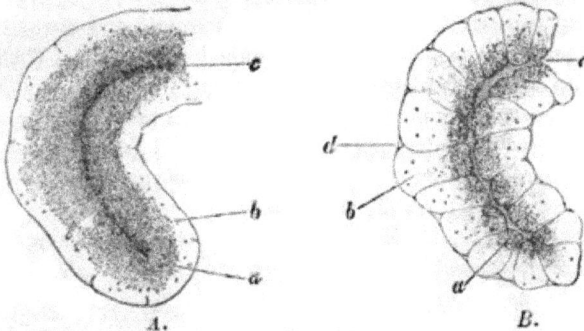

One saccule of the pancreas of the rabbit in different states of activity. *A.* After a period of rest, in which case the outlines of the cells are indistinct, and the inner zone, *i. e.*, the part of the cells (*a*) next the lumen (*c*), is broad and filled with fine granules. *B.* After the gland has poured out its secretion, when the cell outlines (*d*) are clearer, the granular zone (*a*) is smaller, and the clear outer zone is wider.—*From Yeo's Text-Book of Physiology, after Kühne and Lea.*

The pancreatic juice is transparent, colorless, strongly alkaline and viscid, and has a specific gravity of 1.040. It is one of the most important of the digestive fluids, as it exerts a transforming influence upon all the classes of alimentary principles, and has been shown to contain at least three distinct ferments. It has the following composition :—

COMPOSITION OF PANCREATIC JUICE.

Water	900.76
Albuminoid substances	90.44
Inorganic-salts	8.80
	1000.00

The pancreatic juice is characterized by its action : 1st. Upon *starch*.

When starch is subjected to the action of the juice, it is at once transformed into glucose; the change takes place more rapidly than when saliva is added. This action is caused by the presence of a special ferment, *amylopsin*. 2d. Upon *albumen*. The albuminous bodies are changed by the juice into, first, an alkali albumen, and then into peptone. The albumen does not swell up, as is the case in gastric digestion, but is gradually corroded and dissolved. This change is due to the presence of the ferment, *trypsin*. Long-continued action of trypsin converts the peptones into two crystalline bodies, leucine and tyrosin. 3d. Upon *fats*. The most striking action of the pancreatic juice is the *emulsification* of the fats or their subdivision into minute particles of microscopic size. This change takes place rapidly and depends upon the alkalinity of the fluid and the quantity of albumen present, combined with the intestinal movements. The *neutral fats* are also decomposed into their corresponding *fatty acids* and *glycerine ;* the acids thus set free unite with the alkaline bases present in the intestine and form soaps. This decomposition of the neutral fats is caused by the ferment, *steapsin*. 4th. Upon *cane sugar* the juice also exerts a special influence, converting it readily into glucose.

The total quantity of this fluid secreted in twenty-four hours has not been accurately determined; it varies from one to two pounds; it is poured out most abundantly an hour after meals.

The **Bile** has an important influence in the elaboration of the food and its preparation for absorption. It is a golden-brown, viscid fluid, having a neutral or slightly alkaline reaction and a specific gravity of 1.020.

COMPOSITION OF BILE.

Water	859.2
Sodium glycocholate } Sodium taurocholate }	91.4
Fat	9.2
Cholesterine	2.6
Mucus and coloring matter	29.8
Salts	7.8
	1000.00

The *Biliary salts*, sodium glycocholate and taurocholate, are characteristic ingredients, and are formed in the liver by the process of secretion, from materials furnished by the blood. It is probable that they are derived from the nitrogenized compounds, though the stages in the process are unknown. They are reabsorbed from the small intestine to play some ulterior part in nutrition.

Cholesterine is a product of waste taken up by the blood from the nervous tissues and excreted by the liver. It crystallizes in the form of rhombic plates, which are quite transparent. When retained within the blood, it gives rise to the condition of *cholesteræmia*, attended with severe nervous symptoms. It is given off in the fæces under the form of stercorine.

The *Coloring matters* which give the tints to the bile are *biliverdin* and *bilirubin*, and are probably derived from the coloring matter of the blood. Their presence in any fluid can be recognized by adding to it nitric acid containing nitrous acid, when a play of colors is observed, beginning with green, blue, violet, red and yellow.

The Bile is both a secretion and an excretion; it is constantly being formed and discharged by the hepatic ducts into the gall bladder, in which it is stored up, during the intervals of digestion. As soon as food enters the intestines, it is poured out abundantly, by the contraction of the walls of the gall bladder.

The *Amount* secreted in 24 hours is about 2½ pounds.

Functions of the Bile. (1) It assists in the *emulsification* of the fats and promotes their absorption. (2) It tends to prevent putrefactive changes in the food. (3) It stimulates the secretions of the intestinal glands, and excites the normal peristaltic movements of the bowels.

The digested food, the *chyme*, is a grayish, pultaceous mass, but as it passes through the intestines it becomes yellow, from admixture with the bile. It is propelled onward by vermicular motion; by the contraction of the circular and longitudinal muscular fibres.

As the digested food passes through the intestines, the nutritious matters are absorbed into the blood, and the residue enters the large intestine.

The Fæces consist chiefly of indigestible matters, *excretin*, *stercorin* and salts; varying in amount from 4 to 7 oz. in 24 hours.

Defecation is the voluntary act of extruding the fæces from the body; accomplished by a relaxation of the sphincter muscle, the contraction of the walls of the rectum, assisted by the abdominal muscles.

The Gases contained in the stomach and small intestine are oxygen, nitrogen, hydrogen and carbonic acid. In the large intestine, carbonic acid, sulphuretted and carburetted hydrogen. They are introduced with the food, and also developed by chemical changes in the alimentary canal. They distend the intestines, aid capillary circulation, and tend to prevent pressure.

ABSORPTION.

Absorption has for its object the introduction of new materials into the blood, and takes place mainly from the alimentary tract, but also to some extent from the skin, respiratory surface and closed cavities of the body.

The *Agents of Absorption* are the veins and lymphatics.

As a result of the process of digestion, the different alimentary substances

FIG. 5.

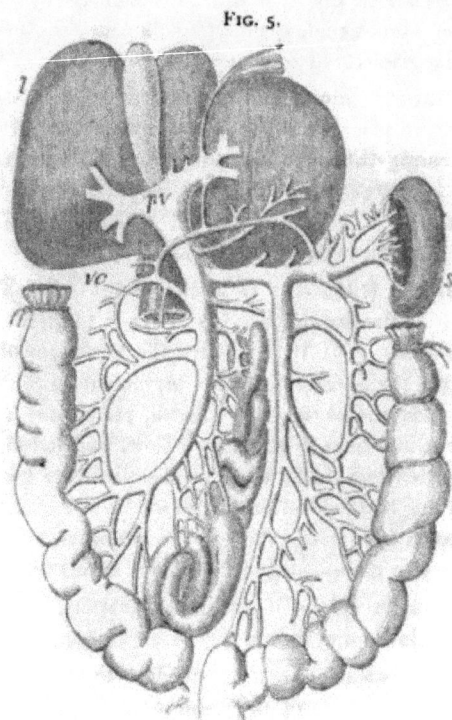

Diagram of the portal vein (*pv*) arising in the alimentary tract and spleen (*s*), and carrying the blood from these organs to the liver.—*From Yeo's Text-Book of Physiology.*

are converted into forms which are capable of being absorbed into the blood, *e. g.*, *albuminose, glucose and fatty emulsion,* water and inorganic matters undergoing no change, being already in a condition to be absorbed and to play a part in the nutritive process.

The blood vessels which are most active as absorbents, are the *gastric, superior* and *inferior mesenteric veins.* They arise in the coats of the alimentary canal, and as they converge, unite with the splenic vein to form the *portal vein*, which enters the liver (Fig. 5).

As the digested mass of food, the *chyme*, passes through the alimentary canal, a large portion of it disappears; the veins absorb *water, albuminose, glucose, and inorganic salts*, and convey them directly into the liver; the blood of the portal vein being especially rich in these substances.

At times, after the ingestion of large quantities of oleaginous food, the blood vessels take up, in addition, a certain quantity of fatty matter; but this is not usually the case; the *fats* being absorbed by special vessels, the *lymphatics* or *lacteals*.

General Anatomy of the Lymphatic System. The lymphatics constitute a system of minute, delicate, transparent vessels, which, having their origin at the periphery of the body, pass forward toward the centre and empty into the veins at the base of the neck, by means of the thoracic duct. In their course they pass through small ovoid bodies, the *lymphatic glands*.

Origin of the Lymphatics. The lymphatic vessels arise in several distinct ways : 1. In *lymph spaces* or *juice canals*. Throughout the connective tissues of the body are found numbers of small, irregular, stellate spaces, which communicate freely with each other, and represent the ultimate radicles of the lymphatic vessels. They frequently contain lymph corpuscles. These spaces communicate with the lymph trunks, through the medium of the plexuses of lymph capillaries, which are much larger than capillary blood vessels, and are lined with endothelial cells with sinuous margins. 2. In *openings on the surface of serous membranes*. The surfaces of the serous membranes are covered with a layer of endothelial cells. At intervals between these cells are found openings—the *stomata*. These stomata communicate directly, through short canals, with the lymph capillaries. The serous cavities may be thus regarded as large lymph spaces. 3. In *perivascular lymph spaces*. In the brain and spinal cord the capillary blood vessels are surrounded by a sheath, formed of endothelial cells, and which contains lymph. This space is in free communication with the lymphatics. By this means the blood vessel is bathed by the lymph stream.

The lymphatic vessels of the small intestine (*the lacteals*) arise within the villous processes which project from the inner surface of the intestine throughout its entire extent. The wall of the villus is formed by an elevation of the basement membrane, and covered by a layer of columnar epithelial cells. The basis of the villus consists of adenoid tissue, fine plexus of blood vessels, unstriped muscular fibres and the lacteal vessel. The adenoid tissue consists of a number of intercommunicating spaces, containing leucocytes. The lacteal vessel possesses a thin, but distinct wall, composed of endothelial plates, with here and there openings, which bring the interior of the villus into communication with the spaces of the adenoid tissue.

The *structure* of the larger vessels resembles that of the veins, consisting of three coats—

1. *External,* composed of fibrous tissue and muscular fibres, arranged longitudinally. 2. *Middle,* consisting of white fibrous and yellow elastic tissue, non-striated muscular fibres, arranged transversely. 3. *Internal,* composed of an elastic membrane, lined by endothelial cells.

Throughout their course are found numerous *semilunar valves,* looking toward the larger vessels, formed by a folding of the inner coat and strengthened by connective tissue.

Lymphatic Glands consist of an external fibrous covering, from the inner surface of which partitions of fibrous tissue, *the trabeculæ,* pass into the substance of the gland, forming a *stroma* or network, in the meshes of which are found the true *lymph corpuscles.*

The lymphatics which enter the gland are called the *afferent* vessels; those which leave it, the *efferent* vessels.

The **Thoracic Duct** is the general trunk of the lymphatic system, into which the vessels of the lower extremities, of the abdominal organs, of the left side of the head and left arm empty their contents. It is about twenty inches in length, arises in the abdomen, opposite the third lumbar vertebra, by a dilatation, the *receptaculum chyli;* ascends along the vertebral column to the seventh cervical vertebra, and terminates in the venous system at the junction of the internal jugular and subclavian veins on the left side. The lymphatics of the right side of the head, of the right arm and the right side of the thorax, terminate in the *right thoracic duct,* about one inch in length, which joins the venous system at the junction of the internal jugular and subclavian on the right side.

Lymph is a clear, transparent fluid, slightly alkaline, having a saline taste and a specific gravity of 1.022. It is found in the lymphatic vessels throughout the body.

Lymph contains a number of corpuscles (the *leucocytes*) resembling the white corpuscles of the blood, which increase in number as it passes through the lymphatic glands. They are about $\frac{1}{2500}$ of an inch in diameter and somewhat granular; they are discharged into the blood, but their function is obscure. When withdrawn from the vessels, lymph undergoes spontaneous coagulation, separating into serum and clot, as in the case of the blood.

COMPOSITION OF LYMPH.

DR. OWEN REES.

Water	96.536
Proteïds (serum-albumen, fibrin, globulin)	1.320
Extractives (urea, sugar, cholesterine)	1.559
Fatty matter	a trace
Salts	0.585
	100.000

Origin of Lymph. Lymph is undoubtedly a transudation from the capillary blood vessels, occurring during the process of nutrition, and is identical, for the most part, with the liquor sanguinis, or plasma. As new material is constantly exuded, the old is absorbed by the lymphatics, and returned again to the circulation.

Excrementitious matters, as *urea*, *cholesterine*, etc., are also taken up from the tissues by the lymphatics and emptied into the blood.

The *total quantity* of lymph poured into the thoracic duct in 24 hours has been estimated at 3½ lbs.

Chyle. As a result of the process of digestion, the oleaginous matters which have been acted upon by the *pancreatic juice* and *bile* are transformed into a condition of *emulsion*, forming an opaque, milky fluid, termed *chyle*, which adheres to the folds of the mucous membrane and villi.

The Molecules of the fat are first absorbed by the epithelial cells, upon the surface of the villi, through which they pass and enter the lymphatics.

Absorption by the Lacteals. The *lacteals*, or lymphatics of the small intestine, have their origin in the interior of the villi, from which they emerge and form a lymphatic plexus; the larger branches of which pass through the layers of the mesentery, and finally terminate in the thoracic duct (Fig. 6).

• In *the intervals* of digestion the lacteals contain clear, transparent lymph, and are invisible on account of their small size and delicacy. But *during digestion* these vessels become filled, from absorption of the chyle, and form a visible network of white vessels ramifying through the mesentery, and converging toward the *receptaculum chyli*.

The lacteal vessels also absorb a small quantity of *water, albuminose, glucose and salts*.

COMPOSITION OF CHYLE.

Water	902.37
Albumen	35.16
Fibrin	3.70
Extractives	15.65
Fatty matters	36.01
Salts	7.11
	1000.00

The *Products of digestion* find their way into the general circulation by two routes :—

1. *Water, albuminose, glucose and salts* are mainly absorbed by the *gas-*

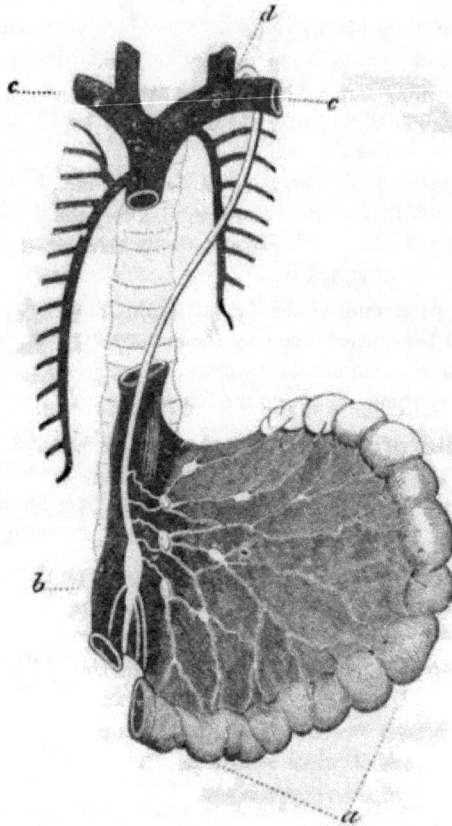

FIG. 6.

Diagram showing the course of the lacteals through the mesentery, and their termination in the thoracic duct. The course of the thoracic duct and its termination at the junction of the internal jugular and subclavian veins.

tric and mesenteric veins, carried into the liver, through the capillaries of which they pass, and enter the inferior vena cava by the hepatic veins.

2. The *Fats* are absorbed by the lacteals, emptied into the thoracic duct, and enter the blood at the junction of the internal jugular and subclavian veins.

Forces aiding the movements of Lymph and Chyle.

1. *Endosmosis.* The continued transudation of matter from the capillaries, and its absorption into the lymphatics by *endosmosis*, constitutes the main cause, the *vis-a-tergo*, of the movement of the lymph; it is so considerable as to rupture the walls of the vessels if they are ligated.

2. *Contraction* of the *non-striated muscular* fibres in the walls of the lymphatic vessels, especially when fully distended, aided by the action of the valves, promotes the onward flow of the fluids.

3. *Muscular contraction* in all parts of the body, by exerting intermittent pressure upon the lymphatic vessels, hastens the current onward; regurgitation being prevented by the closure of the valves.

4. *The inspiratory movement*, by expanding the chest, causes a dilation of the thoracic duct, and a rapid flow of lymph and chyle into it; during expiration it is compressed, and the fluids forcibly expelled into the venous system.

BLOOD.

The Blood is a nutritive fluid containing all the elements necessary for the repair of the tissues; it also contains principles of waste absorbed from the tissues, which are conveyed to the various excretory organs and by them eliminated from the body.

The *total amount* of blood in the body is estimated to be about one-eighth of the body weight; from 16 to 18 pounds in an individual of average physical development. The quantity varies during the 24 hours; the *maximum* being reached in the afternoon, the *minimum* in the early morning hours.

Blood is a heterogeneous, opaque red fluid, having an alkaline reaction, a saline taste, and a specific gravity of 1.055.

The *opacity* is due to the refraction of the rays of light by the elements of which the blood is composed. The *color* varies in hue, from a bright scarlet in the arteries to a deep purple in the veins, due to the presence of a coloring matter, *hæmoglobin*, in different degrees of oxidation.

The *alkalinity* is constant, and depends upon the presence of the alkaline sodium phosphate, $Na_2 H P O_4$.

The *saline taste* is due to the amount of sodium chloride present.

The *specific gravity* ranges within the limits of health, from 1.045 to 1.075.

D

The *odor* of the blood is characteristic, and varies with the animal from which it is drawn, due to the presence of caproic acid.

The *temperature* of the blood ranges from 98° Fahr. at the surface to 107° Fahr. in the hepatic vein; it loses heat by radiation and evaporation as it approaches the extremities, and as it passes through the lungs.

Blood consists of two portions :—

1. The *Liquor Sanguinis* or *Plasma*, a transparent, colorless fluid, in which are floating—

2. *Red and white corpuscles ;* these constituting by weight less than one-half, 40 per cent., of the entire amount of blood.

COMPOSITION OF PLASMA.

	DALTON.
Water	902.00
Albumen	53.00
Paraglobulin	22.00
Fibrinogen	3.00
Fatty matters	2.50
Crystallizable nitrogenous matters	4.00
Other organic matter	5.00
Mineral salts	8.50
	1000.00

Water acts as a solvent for the inorganic matters and holds in suspension the corpuscular elements.

Albumen is the nutritious principle of the blood; it is absorbed by the tissues to repair their waste and is transformed into the organic basis charactertistic of each structure.

Paraglobulin or *fibrinoplastin* is a soft amorphous substance precipitated by sodium chloride in excess, or by passing a stream of carbonic acid through dilute serum.

Fibrinogen can also be obtained by *strongly diluting* the serum and passing carbonic acid through it for a long time, when it is precipitated as a viscous deposit.

Fatty matters exist in small proportion, except in pathological conditions and after the ingestion of food rich in oleaginous matters; it soon disappears, undergoing oxidation, generating heat and force, or is deposited as adipose tissue.

Sugar is represented by glucose, a product of the digestion of saccharine matter and starches in the alimentary canal; glycogenic matter is derived from the liver.

The *Saline constituents* aid the process of osmosis, give alkalinity to the blood, promote the absorption of carbonic acid from the tissues into the blood, and hold other substances in solution; the most important are the sodium and potassium chlorides, the calcium and magnesium phosphates.

Excrementitious matters are represented by carbonic acid, urea, creatin, creatinin, urates, oxalates, etc.; they are absorbed from the tissues by the blood and conveyed to the excretory organs, lungs, kidneys, etc.

Gases. Oxygen, nitrogen, and carbonic acid exist in varying proportions.

BLOOD CORPUSCLES.

The corpuscular elements of the blood occur under two distinct forms, which, from their color, are known as the *red* and *white* corpuscles.

The *Red Corpuscles*, as they float in a thin layer of the Liquor Sanguinis, are of a pale straw color; it is only when aggregated in masses that they assume the bright red color. In form they are circular and biconcave; they have an average diameter of the $\frac{1}{3200}$ of an inch.

In mammals, birds, reptiles, amphibia and fish the corpuscles vary in size and number, gradually becoming larger and less numerous as the scale of animal life is descended, *e. g.* :—

TABLE SHOWING COMPARATIVE DIAMETER OF RED CORPUSCLES.

Mammals.		Birds		Reptiles.		Amphibia.		Fish.	
Man,	$\frac{1}{3200}$	Eagle,	$\frac{1}{1615}$	Turtle,	$\frac{1}{1031}$	Frog,	$\frac{1}{1108}$	Perch,	$\frac{1}{2833}$
Chimpanzee,	$\frac{1}{3212}$	Owl,	$\frac{1}{1762}$	Tortoise,	$\frac{1}{1252}$	Toad,	$\frac{1}{1012}$	Carp,	$\frac{1}{3141}$
Ourang,	$\frac{1}{3289}$	Sparrow,	$\frac{1}{2146}$	Lizard,	$\frac{1}{1515}$	Proteus,	$\frac{1}{460}$	Pike,	$\frac{1}{2060}$
Dog,	$\frac{1}{3542}$	Swallow,	$\frac{1}{2123}$	Viper,	$\frac{1}{1274}$	Siren,	$\frac{1}{470}$	Eel,	$\frac{1}{1742}$
Cat,	$\frac{1}{4404}$	Pigeon,	$\frac{1}{1673}$			Amphiuma,	$\frac{1}{363}$		
Hog,	$\frac{1}{4230}$	Turkey,	$\frac{1}{2643}$						
Horse,	$\frac{1}{4603}$	Goose,	$\frac{1}{1886}$						
Ox,	$\frac{1}{4267}$	Swan,	$\frac{1}{1803}$						

In man and the mammals the red corpuscles present neither a nucleus nor a cell wall, and are universally of a small size. They can be readily distinguished from the corpuscles of birds, reptiles and fish, in which they are larger, oval in shape and possess a well-defined nucleus.

The red corpuscles are exceedingly numerous, amounting to about 5,000,000 in a cubic millimetre of blood. In structure they consist of a firm, elastic, colorless framework, the *stroma*, in the meshes of which is entangled the coloring matter, the *hæmoglobin*.

CHEMICAL COMPOSITION OF RED CORPUSCLES.

Water	688.00
Globulin	282.22
Hæmoglobin	16.75
Fatty matter	2.31
Extractives	2.60
Mineral salts	8.12
	1000.00

Hæmoglobin, the coloring matter of the corpuscles, is an albuminous compound, composed of C. O. H. N. S. and iron. It may exist either in an amorphous or crystalline form. When deprived of all its oxygen, except the quantity entering into its intimate composition, the hæmoglobin becomes darker in color, somewhat purple in hue, and is known as *reducea hæmoglobin*. When exposed to the action of oxygen, it again absorbs a definite amount and becomes scarlet in color, and is known as *oxy-hæmoglobin*. The amount of oxygen absorbed is 1.76 c.cm. ($\frac{7}{10}$ cubic inch) for 1 milligramme ($\frac{1}{64}$) grain of hæmoglobin.

It is this substance which gives the color to the venous and arterial blood. As the venous blood passes through the capillaries of the lungs, the *reduced hæmoglobin* absorbs the oxygen from the pulmonary air and becomes *oxy-hæmoglobin*, scarlet in color, and the blood becomes arterial. When the arterial blood passes into the systemic capillaries, the oxygen is absorbed by the tissues, the hæmoglobin becomes reduced, purple in color, and the blood becomes venous. A dilute solution of oxy-hæmoglobin gives two absorption bands between the lines D and E of the solar spectrum. Reduced hæmoglobin gives but one absorption band, occupying the space existing between the two bands of the oxy-hæmoglobin spectrum.

The *Function* of the red corpuscles is, therefore, to absorb oxygen and carry it to the tissues; the smaller the corpuscles, and the greater the number, the greater is the quantity of oxygen absorbed; and, consequently, all the vital functions of the body become more active.

The *White Corpuscles* are far less numerous than the red, the proportion being, on an average, about 1 white to 350 or 400 red; they are globular in shape, and measure the $\frac{1}{2500}$ of an inch in diameter, and consist of a soft, granular, colorless substance, containing several nuclei.

The *white corpuscles* possess the power of spontaneous movement, alternately contracting and expanding, throwing out processes of their substance and quickly withdrawing them, thus changing their shape from moment to moment. These movements resemble those of the amœba, and for this reason are termed *amœboid*. They also possess the capability of moving from place to place. In the interior of the vessels they adhere to the inner surface, while the red corpuscles move through the centre of the stream.

The white corpuscles are identical with the leucocytes, and are found in milk, lymph, chyle and other fluids.

Origin of Corpuscles. The red corpuscles take their origin from the mesoblastic cells in the vascular area of the developing embryo.

In the adult they are produced from colorless nucleated corpuscles resembling the white corpuscles. The spleen is the organ in which they are finally destroyed.

The white corpuscles originate from the leucocytes of the adenoid tissue, and subsequently give rise to the red corpuscles and partly to new tissues that result from inflammatory action.

COAGULATION OF THE BLOOD.

When blood is withdrawn from the body and allowed to remain at rest, it becomes somewhat thick and viscid in from three to five minutes; this viscidity gradually increases until the entire volume of blood assumes a jelly-like consistence, which occupies from five to fifteen minutes.

As soon as coagulation is completed, a second process begins, which consists in the contraction of the coagulum and the oozing of a clear, straw-colored liquid, the *serum*, which gradually increases in quantity as the *clot* diminishes in size, by contraction, until the separation is completed, which occupies from 12 to 24 hours.

The changes in the blood are as follows :—

Before coagulation.

Living blood. { Liq. Sanguinis or Plasma. } Consisting of { Water. Albumen. Fibrinogen. Salts.

Corpuscles. Red and white.

After coagulation.

Dead blood. { Crassamentum. Clot or coagulum. } Containing { Fibrin. Corpuscles.

Serum. Containing { Water. Albumen. Salts.

The serum, therefore, differs from the Liquor Sanguinis in not containing fibrin.

In from 12 to 24 hours the upper surface of the clot presents a grayish appearance, the *buffy coat*, which is due to the rapid sinking of the red corpuscles beneath the surface, permitting the fibrin to coagulate without them, which then assumes a grayish-yellow tint. Inasmuch as the white corpuscles possess a lighter specific gravity than the red, they do not sink so rapidly, and becoming entangled in the fibrin, assist in forming the buffy coat. Continued contraction gives a *cupped* appearance to the surface of the clot.

Inflammatory states of the blood produce a marked increase in the

buffed and cupped condition, on account of the aggregation of the corpuscles, and their tendency to rapid sinking.

Nature of Coagulation. Coagulated fibrin does not preëxist in the
blood, but is formed at the moment blood is withdrawn from the vessels.
According to Denis, a liquid substance, *plasmine*, exists in the blood,
which, when withdrawn from the circulation, decomposes into *fibrin* and
met-albumen.

According to Schmidt, fibrin results from the union of *fibrinoplastin*
(paraglobulin) and *fibrinogen*, brought about by the presence of a third
substance, the fibrin ferment.

According to Hammersten and others, the fibrin obtained from the
blood after coagulation, comes from the fibrinogen alone, the conversion
being brought about by the presence of a ferment substance, paraglobulin
in this case having nothing to do with the change. This view is supported
by the fact that the quantity of fibrin obtained from the blood is never
greater than the quantity of fibrinogen previously present. The origin of
the ferment is obscure, but there is reason to believe that it comes from the
injured vascular coats or from the breaking up of the white corpuscles.

Conditions Influencing Coagulation. The process is *retarded* by
cold, retention within living vessels, neutral salts in excess, inflammatory
conditions of the system, imperfect aeration, exclusion from air, etc.

It is *hastened* by a temperature of 100° F., contact with air, rough surfaces and rest.

Blood coagulates in the body after the arrest of the circulation in the
course of 12 to 24 hours; local arrest of the circulation, from compression
or a ligature, will cause coagulation, thus preventing hemorrhages from
wounded vessels.

The Composition of the Blood varies in different portions of the
body. The *arterial* differs from the *venous*, in being more coagulable, in
containing more oxygen and less carbonic acid, in having a bright scarlet
color, from the union of oxygen with hæmoglobin; the purple hue of
venous blood results from the deoxidation of the coloring matter.

The blood of the *portal vein* differs in constitution, according to different
stages of the digestive process; during digestion it is richer in water,
albuminous matter and sugar; occasionally it contains fat; corpuscles are
diminished, and there is an absence of biliary substances.

The blood of the *hepatic vein* contains a larger proportion of red and
white corpuscles; the sugar is augmented, while albumen, fat and fibrin
are diminished.

Pathological conditions of the Blood.

1. *Plethora*—increase in the volume or quantity of blood.

2. *Anæmia*—deficiency of red globules with increase of water.

3. *Leucocythemia*—increase of white and diminution of red corpuscles.

4. *Glycohæmia*—excess of sugar in the blood.

5. *Uræmia*—increase in the amount of urea.

6. *Cholesteræmia*—an excess of cholesterine in the blood.

7. *Thrombosis* and *embolism*—clotting of blood in the **vessels and** dissemination **of coagula.**

8. *Lipæmia*—**an excess of fat.**

9. *Melanæmia*—pigment in the blood.

CIRCULATION OF THE BLOOD.

The Object of the Circulation is to distribute nutritious blood to all portions of the system and to carry waste materials from the tissues to the various eliminating organs.

The Circulatory Apparatus consists of the heart, arteries, capillaries and veins.

The Heart is a hollow, muscular organ, pyramidal in shape, measuring 5½ inches in length and weighing from 10 to 12 ozs. in the male, and 8 to 10 ozs. in the female. It is invested externally by a closed fibro-serous sac, the *pericardium*, containing a small amount of fluid, which prevents friction as the visceral and parietal layers glide over each other, during the movements of the heart and lungs.

The heart consists of four cavities, a right auricle and ventricle, and a left **auricle and ventricle,** completely separated by a vertical partition. The *right* is the *venous side*, receiving the blood from the venæ cavæ, and propelling it through the pulmonary artery into the lungs; the *left* is the *arterial side*, receiving **the arterial** blood from the lungs by the pulmonary veins, and propelling it through the aorta to the system at large.

The *Auriculo-ventricular orifices* are guarded on the right and left sides by the *tricuspid* and *mitral* valves respectively, while they are so arranged as to permit the flow of blood in the forward direction only; the orifices of the *pulmonary artery* and *aorta* are guarded by the *semilunar* valves.

The *Endocardium* is a delicate, shining membrane, lining the interior of the heart, and continuous with the lining membrane of the blood vessels.

The walls of the left ventricle are nearly half an inch in diameter, being **two or three** times thicker than the walls of the right; the force of its **contraction is, therefore,** much greater.

The Function of the Heart is to propel the blood to all portions of the vascular system; accomplished by successive alternate contractions and relaxations of its muscular walls, constituting the *systole* and *diastole*.

Course of the blood through the Heart. The venous blood returned to the heart by the superior and inferior venæ cavæ is emptied, during the diastole, into the right auricle, on the contraction of which it is forced through the right auriculo-ventricular opening into the right ventricle and distends it. Upon contraction of the ventricle, the blood is propelled through the pulmonary artery into the lungs, where it undergoes aeration and is changed in color.

The arterial blood is now collected by the pulmonary veins and poured into the left auricle; thence it passes into the left ventricle, which becomes fully distended. Upon the contraction of the ventricle, the blood is propelled into the aorta, and by it distributed to the system at large, to be again returned to the heart by the veins.

Regurgitation from the ventricles into the auricles during the systole is prevented by the closure of the tricuspid and mitral valves; regurgitation from the pulmonary artery and aorta into the ventricles during the diastole is prevented by the closure of the semilunar valves.

Movements of the Heart. At each revolution, during the systole, the heart hardens and becomes shortened in its long diameter; its apex is raised up, rotated on its axis from left to right and thrown forward against the walls of the chest. The *impulse* of the heart, observed about two inches below the nipple, and one inch to the sternal side, between the fifth and sixth ribs, is caused mainly by the apex of the heart striking against the chest walls, assisted by the distention of the great vessels about the base of the heart.

Sounds of the Heart. If the ear be placed over the cardiac region, two distinct sounds are heard during each revolution of the heart, closely following each other and which differ in character.

The sound coinciding with the systole in point of time, the *first sound*, is long and dull, and caused by the closure and vibration of the auriculo-ventricular valves, the contraction of the walls of the ventricles and the apex beat; the *second sound*, occurring during the *diastole*, is short and sharp, and caused by the closure of the semilunar valves.

The *capacity* of the left ventricle when fully distended is estimated at from four to seven ounces.

The frequency of the heart's action varies at different periods of life, but in the adult male it beats about 72 times per minute. It is influenced by age, exercise, posture, digestion, etc.

Age. Before birth, the *number of pulsations* per minute averages 140
During the first year it diminishes to 128
During the third year diminishes to.......... 95
From the eighth to the fourteenth years averages.... 84
In adult life the average is...................................... 72

Exercise and *digestion* increase the frequency of the heart's action.

Posture influences the number of pulsations per minute ; in the male, standing, the average is 81 ; sitting, 71 ; lying, 66 ; independent, for the most part, of muscular effort.

The Rhythmical movements of the heart **are dependent upon—**1. An inherent irritability of the muscular fibre, which **manifests itself as long as** the nutrition is maintained. 2. The continuous flow **of blood through** its cavities, distending them and stimulating the endocardium.

The **force exerted by the left ventricle at** each contraction has been estimated at 52 pounds. If a tube be inserted into the aorta, the pressure there will be sufficient to support a column of blood nine feet or a column of mercury six inches in height, the weight in either case being about four pounds. **The** estimation of the force which the heart is required to exert to support this column of blood, is arrived at by multiplying the pressure in the **aorta (4** pounds) by the area **of** the internal surface of the left ventricle (about 13 inches). Each **inch of the** ventricle being capable of supporting a downward pressure of **4 pounds.**

Work done by the Heart. The work done by the heart is estimated by **multiplying** the amount of blood sent out from the right and left ventricles at each contraction, by the pressure in the pulmonary artery and aorta respectively, *e. g.*, when the **right** ventricle **contracts,** it forces out one-quarter pound of blood, **and in so** doing must overcome a pressure in the pulmonary artery **sufficient to support** a column of blood three feet in height; that is, **must exert** energy sufficient to raise ¼ lb. 3 feet, or ¼ × 3 or ¾ **lb. one foot. When the left** ventricle contracts, it sends out ¼ lb. of blood, **and in so doing,** the left ventricle must overcome a pressure in the aorta sufficient to support a column of blood nine feet in height; that is must exert energy sufficient to raise ¼ lb. 9 feet, or ¼ × 9 or 2¼ lbs. one foot. Work done is estimated by the amount of energy required to raise a definite weight a definite height, the unit, the foot pound, being that required to raise one pound one foot.

The heart, therefore, at each systole exerts energy sufficient to raise 3 foot **pounds, and as it** contracts 72 times per minute, it would raise in that **time 3 × 72 or** 216 foot pounds ; and in one hour 216 × 60 or 12,960 foot pounds; and in 24 hours 12,960 × 24 or 311,040 foot pounds or 138.5 foot tons.

Influence of the Nervous System upon the Heart. When the heart of a frog is removed from the body, it continues to beat for a variable length of time, depending upon the nature of the conditions surrounding it. The heart of warm-blooded animals continues to beat but for a very short time. The cause of the continued pulsations of the frog heart is the presence of nervous ganglia in its substance. These ganglia have not been shown to exist in the mammalian heart, but there is reason to believe that the nervous mechanism is fundamentally the same.

The ganglia of the heart are three in number, one situated at the opening of the inferior vena cava (the ganglion of Remak), a second situated in the auriculo-ventricular septum (the ganglion of Bidder), and a third situated in the inter-auricular septum (the ganglion of Ludwig). The first two are motor in function and excite the pulsations of the heart; the third is inhibitory in function and retards the action of the heart. The actions of these ganglia, though for the most part automatic, are modified by impressions coming through nerves from the medulla oblongata. When the inhibitory centre is stimulated by muscarin, the heart is arrested in diastole; when atropia is applied, the heart recommences to beat, because atropia paralyzes the inhibitory centre.

The nerves modifying the action of the heart are the **Pneumogastric** (Vagus) and the Accelerator nerves.

The *Pneumogastric nerve*, after emerging from the medulla, receives motor fibres from the spinal accessory nerve. It then passes downward, giving off branches, some of which terminate in the inhibitory ganglion. *Stimulation* of the vagus by increasing the activity of the inhibitory centre arrests the heart in diastole with its cavities full of blood; but as the stimulation is only temporary, after a few seconds the heart recommences to beat; at first the pulsations are weak and feeble, but soon regain their original vigor. After the administration of atropia in sufficient doses to destroy the termination of the pneumogastric, stimulation of its trunk has no effect upon the heart. The inhibitory fibres in the vagus are constantly in action, for *division* of the nerve on both sides is always followed by an increase in the frequency of the heart's pulsations.

The *Accelerator fibres* arise in the medulla, pass down the cord, emerge in the cervical region, pass to the last cervical and first dorsal ganglia of the sympathetic, and thence to the heart. *Stimulation* of these fibres causes an increased frequency of the heart's pulsations, but they are diminished in force.

ARTERIES.

The Arteries are a series of branching tubes conveying blood to all portions of the body. They are composed of three coats—

1. *External*, formed of areolar and elastic tissue.
2. *Middle*, contains both elastic and muscular fibres, arranged transversely to the long axis of the artery. The elastic tissue is more abundant in the larger vessels, the muscular in the smaller.
3. *Internal*, composed of a thin homogeneous membrane, covered with a layer of elongated endothelial cells.

The arteries possess both *elasticity* and *contractility*.

The Property of Elasticity allows the arteries already full to accommodate themselves to the incoming **amount** of blood, and to convert the intermittent acceleration of blood in the large vessels into a steady and continuous stream in the **capillaries**.

The *Contractility* of the smaller vessels equalizes the current of blood, regulates the amount going to each part, and promotes the onward flow of blood.

Blood Pressure. Under the influence of the ventricular systole, the **recoil of** the elastic walls of the arteries, and the resistance offered by the capillaries, the blood is constantly being subjected to a certain amount of pressure. If a large artery of an animal be divided, and a glass tube of the same calibre be inserted into its orifice, **the blood** will rise to a height of about nine feet ; or if it be connected with a mercurial manometer, the mercury will rise to a height of six inches. This height will be a measure of the pressure in the vessel. The absolute quantity of mercury sustained **by an** artery can be arrived at by multiplying the height of the column by **the area of a transverse** section of that artery.

The pressure of the blood is greatest in the large arteries, but **gradually** decreases toward the capillaries.

The blood pressure is increased or diminished by influences acting **upon** the heart or upon the peripheral resistance of the capillaries, viz. :—

If, while **the force** of the heart remains the same, the number of pulsations per minute increases, thus increasing the volume of blood in the arteries, the pressure rises. If the rate remains the same, but the force increases, the pressure again rises. Causes that increase the peripheral resistance by contracting the arterioles, *e. g.*, vasomotor nerves, cold, etc., produce an increase of the pressure.

On the other hand, influences which *diminish* either the volume of the blood, **or** the number of pulsations, or the force of the heart, or the peripheral resistance, *lower* the pressure.

The Pulse is the sudden distention of the artery in a transverse and longitudinal direction, due to the injection of a volume of blood into the arteries at the time of the ventricular systole. As the vessels are already full of blood, they must expand in order to accommodate themselves to the incoming volume of blood. The blood pressure is thus increased, and the pressure originating at the ventricle excites a *pulse wave*, which passes from the heart toward the capillaries at the rate of about twenty-nine feet per second. It is this wave that is appreciated by the finger.

The Velocity with which the blood flows in the arteries diminishes from the heart to the capillaries, owing to an increase of the united sectional area of the vessels, and increases in rapidity from the capillaries toward the heart. It moves most rapidly in the large vessels, and especially under the influence of the ventricular systole. From experiments on animals, it has been estimated to move in the carotid of man at the rate of sixteen inches per second, and in the large veins at the rate of four inches per second.

The Calibre of the blood vessels is regulated by the *vasomotor* nerves, which have their origin in the gray matter of the medulla oblongata. They issue from the spinal cord through the anterior roots of spinal nerves, pass through the sympathetic ganglia, and ultimately are distributed to the coats of the blood vessels. They exert, at different times, a constricting and dilating action upon the vessels, thus keeping up the arterial tonus.

Capillaries. The capillaries constitute a network of vessels of microscopic size, which distribute the blood to the inmost recesses of the tissues, inosculating with the arteries on the one hand and the veins on the other; they branch and communicate in every possible direction.

The diameter of a capillary vessel varies from the $\frac{1}{5000}$ to the $\frac{1}{3000}$ of an inch; their walls consist of a delicate homogeneous membrane, the $\frac{1}{20000}$ of an inch in thickness, lined by flattened, elongated, endothelial cells, between which, here and there, are observed *stomata*.

It is through the agency of the capillary-vessels that the phenomena of nutrition and secretion takes place, for here the blood flows in an equable and continuous current, and is brought into intimate relationship with the tissues, two of the essential conditions for proper nutrition.

The *rate* of *movement* in the capillary vessels is estimated at one inch in thirty seconds.

In the capillary current the red corpuscles may be seen hurrying down the centre of the stream, while the white corpuscles in the still layer adhere to the walls of the vessel, and at times can be seen to pass through the walls of the vessel by amœboid movements.

The passage of the blood through the capillaries is mainly due to the force of the ventricular systole and the elasticity of the arteries; but it is probably also aided by a power resident in the capillaries themselves, the result of a vital relation between the blood and the tissues.

The Veins are the vessels which return the blood to the heart; they have their origin in the venous radicles, and as they approach the heart, converge to form larger trunks, and terminate finally in the venæ cavæ.

They possess three coats—

1. *External*, made up of areolar tissue.

2. *Middle*, composed of non-striated muscular fibres, yellow, elastic and fibrous tissue.

3. *Internal*, an endothelial membrane, similar to that of the arteries.

Veins are distinguished by the possession of valves throughout their course, which are arranged in pairs, and formed by a reflection of the internal coat, strengthened by fibrous tissues; they always look toward the heart, and when closed prevent a return of blood in the veins. Valves are most numerous in the veins of the extremities, but are entirely absent in many others.

The onward flow of blood in the veins is mainly due to the action of the heart; but is assisted by the contraction of the voluntary muscles and the force of aspiration.

Muscular contraction, which is intermittent, aids the flow of blood in the veins, by compressing them. As regurgitation is prevented by the closure of the valves, the blood is forced onward toward the heart.

Rhythmical movements of veins have been observed in some of the lower animals, aiding the onward current of blood.

During the movement of *inspiration* the thorax is enlarged in all its diameters, and the pressure on its contents at once diminishes. Under these circumstances a suction force is exerted upon the great venous trunks, which causes the blood to flow with increased rapidity and volume toward the heart.

Venous pressure. As the force of the heart is nearly expended in driving the blood through the capillaries, the pressure in the venous system is not very marked, not amounting in the jugular vein of a dog to more than $\frac{1}{12}$ that of the carotid artery.

The time required for a complete circulation of the blood throughout the vascular system has been estimated to be from 20 to 30 seconds, while for the *entire mass* of blood to pass through the heart 58 pulsations would be required, occupying 48 seconds.

The Forces keeping the blood in circulation are—

1. Action of the heart.
2. Elasticity of the arteries.
3. Capillary force.
4. Contraction of the voluntary muscles upon the veins.
5. Respiratory movements.

RESPIRATION.

Respiration is the function by which oxygen is absorbed into the blood and carbonic acid exhaled. The appropriation of the oxygen and the evolution of carbonic acid takes place in the tissues as a part of the general nutritive processs; the blood and respiratory apparatus constituting the media by means of which the interchange of gases is accomplished.

The Respiratory Apparatus consists of the larynx, trachea and lungs.

The Larynx is composed of firm cartilages, united together by ligaments and muscles; running antero-posteriorly across the upper opening are four ligamentous bands, the two superior or *false* vocal cords, and the two inferior or *true* vocal cords, formed by folds of the mucous membrane. They are attached anteriorly to the thyroid cartilages and posteriorly to the arytenoid cartilages, and are capable of being separated by the contraction of the posterior crico-arytenoid muscles, so as to admit the passage of air into and from the lungs.

The Trachea is a tube from four to five inches in length, three-quarters of an inch in diameter, extending from the cricoid cartilage of the larynx to the third dorsal vertebra, where it divides into the right and left bronchi. It is composed of a series of cartilaginous rings, which extend about two-thirds around its circumference, the posterior third being occupied by fibrous tissue and non-striated muscular fibres which are capable of diminishing its calibre.

The trachea is covered externally by a tough, fibro-elastic membrane, and internally by mucous membrane, lined by columnar ciliated epithelial cells. The cilia are always waving from within outward. When the two bronchi enter the lungs they divide and subdivide into numerous and smaller branches, which penetrate the lung in every direction until they finally terminate in the *pulmonary lobules.*

As the bronchial tubes become smaller their walls become thinner; the cartilaginous rings disappear, but are replaced by irregular angular plates of cartilage; when the tube becomes less than the $\frac{1}{50}$ of an inch in diameter they wholly disappear, and the fibrous and mucous coats blend together, forming a delicate, elastic membrane, with circular muscular fibres.

The Lungs occupy the cavity of the thorax, are conical in shape, of a pink color and a spongy texture. They are composed of a great number of distinct lobules, the *pulmonary lobules*, connected together by interlobular connective tissue. These lobules vary in size, are of an oblong shape, and are composed of the ultimate ramifications of the bronchial tubes, within which are contained the *air vesicles or cells*. The walls of the air vesicles, exceedingly thin and delicate, are lined internally by a layer of tessellated epithelium, externally covered by elastic fibres, which give the lungs their elasticity and distensibility.

FIG. 7.

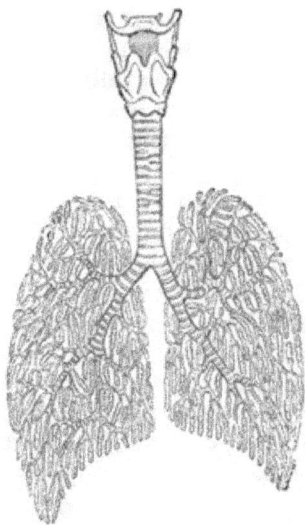

Diagram of the respiratory organs. The windpipe leading down from the larynx is seen to branch into two large bronchi, which subdivide after they enter their respective lungs.

The Venous Blood is distributed to the lungs for aeration by the pulmonary artery, the terminal branches of which form a rich plexus of capillary vessels surrounding the air cells; the air and blood are thus brought into intimate relationship, being separated only by the delicate walls of the air cells and capillaries.

The thoracic cavity in which the respiratory organs are lodged is of a conical shape, having its apex directed upward, its base downward. Its framework is formed posteriorly by the spinal column, anteriorly by the sternum, and laterally by the ribs and costal cartilages. Between and over the ribs lie muscles, fascia and skin; above the thorax is completely closed by the structures passing into it and by the cervical fascia and skin; below it is closed by the diaphragm. It is therefore an air-tight cavity.

The Pleura. Each lung is surrounded by a closed serous membrane, the pleura, one layer of which, the *visceral*, is reflected over the lung, the other, the *parietal*, reflected over the wall of the thorax; between the two layers is a small amount of fluid which prevents friction during the play of the lungs in respiration.

Owing to the elastic tissue which is present in the lungs, they are very readily distensible, so much so, indeed, that the pressure of the air inside the trachea and lungs is sufficient to distend them until they completely fill all parts of the thoracic cavity not occupied by the heart and great vessels. The elastic tissue endows them not only with distensibility, but also with the power of elastic recoil, by which they are enabled to accommodate themselves to all variations in the size of the thoracic cavity.

When the chest walls recede, the air within the lungs expands and presses them against the ribs; when the chest walls contract, the air being driven out, the elastic tissue recoils and the lungs return to their original condition. The movements of the lungs are therefore entirely passive.

As the capacity of the chest in a state of rest is greater than the volume of the lungs after they are collapsed, it is quite evident that in the living condition the lungs are distended and in a state of elastic tension, which is greater or less in proportion as the thoracic cavity is increased or diminished in size. The elastic tissue, always on the stretch, is endeavoring to pull the visceral layer of the pleura away from the parietal layer, but is antagonized by the pressure of the air within the air passages. This condition of things persists as long as the thoracic cavity remains air-tight; but if an opening be made in the thoracic wall, the pressure of the external air which was previously supported by the practically rigid walls of the thorax now presses upon the lung with as much force as the air within the lung. The two pressures being neutralized, there is nothing to prevent the elastic tissue from recoiling, driving the air out and collapsing. The elastic tension of the lungs can be readily measured in man after death by inserting a manometer into the trachea. Upon opening the thorax and allowing the tissue to recoil, the air presses upon the mercury and elevates it, the extent to which it is raised being the index of the pressure. Hutchinson calculated the pressure to be one-half pound to the square inch of the lung surface.

Respiratory movements. The movements of respiration are two, and consist of an alternate dilatation and contraction of the chest, known as inspiration and expiration.

1. *Inspiration* is an active process, the result of the expansion of the thorax, whereby air is introduced into the lungs.

2. *Expiration* is a partially passive process, the result of the recoil of the elastic walls of the thorax, and the recoil of the elastic tissue of the lungs, whereby the carbonic acid is expelled.

In Inspiration the chest is enlarged by an increase in all its diameters, viz.:—

1. The *vertical* is increased by the contraction and descent of the diaphragm when it approximates a straight line.

2. The *antero-posterior* and *transverse* diameters are increased by the elevation and rotation of the ribs upon their axes.

In *ordinary tranquil inspiration* the muscles which elevate the ribs and thrust the sternum forward, and so increase the diameters of the chest, are the *external intercostals,* running from above downward and forward, the *sternal portion* of the *internal intercostals* and the *levatores costarum.*

In the **extraordinary** *efforts* of inspiration certain auxiliary muscles are brought into play, viz.: the *sterno-mastoid, pectorales, serratus magnus,* which increase the capacity of the thorax to its utmost limit.

In Expiration the diameters of the chest are all diminished, viz.:—

1. The *vertical*, by the ascent of the diaphragm.

2. The *antero-posterior*, by a depression of the ribs and sternum.

In *ordinary tranquil expiration* the diameters of the thorax are diminished by the recoil of the elastic tissue of the lungs and the ribs; but in forcible expiration the muscles which depress the ribs and sternum, and thus further diminish the diameter of the chest, are the *internal intercostals,* the *infracostals,* and the *triangularis sterni.*

In the *extraordinary efforts of expiration* certain auxiliary muscles are brought into play, viz.: the *abdominal* and *sacro-lumbalis muscles,* which diminish the capacity of the thorax to its utmost limit.

Expiration is aided by the recoil of the elastic tissue of the lungs and ribs and the pressure of the air.

Movements of the Glottis. At each inspiration the *rima glottidis* is dilated by a separation of the vocal cords, produced by the contraction of the *crico-arytenoid* muscles, so as to freely admit the passage of air into the lungs: in expiration they fall passively together, but do not interfere with the exit of the air from the chest.

Nervous Mechanism of Respiration. The movements of Respiratory muscles, though capable of being modified to a certain extent by efforts of the will, are of an automatic character, and called forth by nervous impulses emanating from the medulla oblongata. The Respiratory centre, the so-called vital point, generates the nerve impulses, which, traveling outward through the phrenic and intercostal nerves, excite contractions of the diaphragm and intercostal muscles respectively. This centre is for the most part automatic in its action, though it is capable of being modified by impulses reflected to it through various sensory nerves.

This centre may be stimulated—

E

1. *Directly*, by the condition of the blood. An increase of carbonic acid or a diminution of oxygen in the blood causes an acceleration of the respiratory movements; the reverse of these conditions causes a diminution of the respiratory movements.

2. *Indirectly*, by reflex action. The medulla may be excited to action through the pneumogastric nerve, by the presence of carbonic acid in the lungs irritating its terminal filaments; through the fifth nerve, by irritation of the terminal branches; and through the nerves of general sensibility. In either case this centre reflects *motor* impulses to the respiratory muscles through the *phrenic, intercostals, inferior laryngeal* and other nerves.

Types of Respiration. The *abdominal type* is most marked in young children, irrespective of sex; the respiratory movements being effected by the diaphragm and abdominal muscles.

In the *superior costal type*, exhibited by the adult female, the respiratory movements are more marked in the upper part of the chest, from the 1st to the 7th ribs, permitting the uterus to ascend in the abdomen during pregnancy without interfering with respiration.

In the *inferior costal type*, manifested by the male, the movements are largely produced by the muscles of the lower portion of the chest, from the 7th rib downward, assisted by the diaphragm.

The respiratory movements vary according to age, sleep and exercise, being most frequent in early life, but averaging 20 per minute in adult life. They are diminished by sleep and increased by exercise. There are about four pulsations of the heart to each respiratory act.

During inspiration *two sounds* are produced; the one, heard in the thorax, in the trachea and larger bronchial tubes, is *tubular* in character; the other, heard in the substance of the lungs, is *vesicular* in character.

AMOUNT OF AIR EXCHANGED IN RESPIRATION, AND CAPACITY OF LUNGS.

The *Tidal* or *breathing volume* of air, that which passes in and out of the lungs at each inspiration and expiration, is estimated at from 20 to 30 cubic inches.

The *Complemental air* is that amount which can be taken into the lungs by a forced inspiration, in addition to the ordinary tidal volume, and amounts to about 110 cubic inches.

The *Reserve air* is that which usually remains in the chest after the ordinary efforts of expiration, but which can be expelled by forcible expiration. The volume of reserve air is about 100 cubic inches.

The *Residual air* is that portion which remains in the chest and cannot be expelled after the most forcible expiratory efforts, and which amounts, according to Dr. Hutchinson, to about 100 cubic inches.

The **Vital Capacity** of the chest indicates the amount of air that can be forcibly expelled from the lungs after the deepest possible inspiration, and is an index of an individual's power of breathing in disease and prolonged severe exercise. The combined amounts of the tidal, the complemental and reserve air, 230 cubic inches, represents the vital capacity of an individual 5 feet 7 inches in height. The vital capacity varies chiefly with stature. It is increased 8 cubic inches for every inch in height above this standard, and diminishes 8 cubic inches for each inch below it.

The **Tidal Volume** of air is carried only into the trachea and larger bronchial tubes by the inspiratory movements. It reaches the deeper portions of the lungs in obedience to the law of diffusion of gases, which is inversely proportionate to the square root of their densities.

The *ciliary action* of the columnar cells lining the bronchial tubes also assists in the interchange of air and carbonic acid.

The *entire volume of air* passing in and out of the thorax in 24 hours is subject to great variation, but can be readily estimated from the tidal volume and the number of respirations per minute. Assuming that an individual takes into the chest 20 cubic inches at each inspiration, and breathes 18 times per minute, in 24 hours there would pass in and out of the lungs 518,400 cubic inches or 300 cubic feet.

Composition of Air: Oxygen, 20.81 parts; nitrogen, 79.19, forming a mechanical mixture in which exist traces of carbonic acid and watery vapor.

The changes in the air effected by respiration are—

Loss of oxygen, to the extent of 5 cubic inches per 100 of air, or 1 in 20.

Gain of carbonic acid, to the extent of 4.66 cubic inches per 100 of air or .93 inch in 20.

Increase of watery vapor and organic matter.

Elevation of temperature.

Increase and at times decrease of nitrogen.

Gain of ammonia.

The *total quantity* of *oxygen* withdrawn from the air and consumed by the body in 24 hours amounts to 15 cubic feet, and can be readily estimated from the amount consumed at each respiration. Assuming that one inch of oxygen remains in the lungs at each respiration, in one hour there are consumed 18 inches, and in 24 hours, 25,920 cubic inches or 15 cubic feet, weighing 18 oz. To obtain this quantity, 300 cubic feet of air are necessary.

The *quantity* of *carbonic acid* exhaled in 24 hours **varies** greatly. It can be estimated in the same way. Assuming that an individual exhales .93 + cubic inch at each respiration, in one hour there are eliminated 1008 cubic inches, and in 24 hours 24,192 cubic inches or 14 cubic feet, containing 7 ozs. of pure carbon.

As oxygen and carbon unite to form an equal volume of carbonic acid **gas,** there disappears daily in the body one cubic foot of oxygen, which in all probability unites with the surplus hydrogen of the food to form water.

The exhalation of carbonic acid is *increased* by muscular exercise; nitrogenous food; tea, coffee and rice; age, and by muscular development; *decreased* by a lowering of temperature; repose; gin and **brandy, and a** dry condition of the air.

Condition of the Gases in the Blood.

Oxygen is absorbed from the lungs into the arterial blood by the coloring matter, *hæmoglobin*, with which it exists in a state of loose combination, and is disengaged during the process of nutrition.

Carbonic acid, arising in the tissues, is absorbed into the blood, in consequence of its alkalinity; where it exists in a state of simple solution and also in a state of feeble combination with the carbonates, soda and potassa, forming the bicarbonates; it is liberated by pneumic acid in the pulmonary tissue.

Nitrogen is simply held in solution in the plasma.

The *amount* of *watery vapor* thrown off from the lungs daily **is** about one pound, with which is mingled organic matter and ammonia.

Changes in the Blood during Respiration.

As the blood passes through the lungs it is changed in *color*, from the dark purple hue of venous blood to the bright scarlet of arterial blood.

The heterogeneous composition of venous blood is exchanged for **the** uniform composition of the arterial.

It gains oxygen and loses carbonic acid.

Its coagulability is increased. Temperature is diminished.

Asphyxia. If the supply of oxygen to the lungs be diminished and the carbonic acid retained in the blood, the normal respiratory movements cease, the condition of asphyxia ensues, which soon terminates in death.

The *phenomena* of asphyxia are, violent spasmodic action of the respiratory muscles, attended by convulsions of the muscles of the extremities, engorgement of the venous system, lividity of the skin, abolition of sensibility and reflex action, and death.

The *cause* of *death* is a paralysis of the heart, from over distention by blood. The passage of the blood through the capillaries is prevented by contraction of the smaller arteries, from irritation of the vasomotor centre. The heart is enfeebled by a want of oxygen and inhibited in its action by the inhibitory centres.

ANIMAL HEAT.

The Functional Activity of all the organs and tissues of the body is attended by the evolution of heat, which is independent, for the most part, of external conditions. Heat is a *necessary condition* for the due performance of all vital actions; though the body constantly loses heat by *radiation* and *evaporation*, it possesses the capability of renewing it and maintaining it at a fixed standard. The *normal temperature* of the body in the adult, as shown by means of a delicate thermometer placed in the axilla, ranges from 97.25° Fahr. to 99.5° Fahr., though the *mean normal* temperature is estimated by Wunderlich at 98.6° Fahr.

The temperature varies in different portions of the body, according to the degree in which oxidation takes place; being the highest in the muscles during exercise, in the brain, blood, liver, etc.

The conditions which produce variations in the normal temperature of the body are : age, period of the day, exercise, food and drink, climate, season and disease.

Age. At birth the temperature of the infant is about 1° F. above that of the adult, but in a few hours falls to 95.5° F., to be followed in the course of 24 hours by a rise to the normal or a degree beyond. During childhood the temperature approaches that of the adult ; in aged persons the temperature remains about the same, though they are not as capable of resisting the depressing effects of external cold as adults. A *diurnal variation* of the temperature occurs from 1.8° F. to 3.6° F. (Jürgensen); the *maximum* occurring late in the afternoon, from 4 to 9 P. M., the *minimum*, early in the morning, from 1 to 7 A. M.

Exercise. The temperature is raised from 1° to 2° F. during active contractions of the muscular masses, and is probably due to the increased activity of chemical changes; a rise beyond this point being prevented by its diffusion to the surface, consequent on a more rapid circulation, radiation, more rapid breathing, etc.

Food and drink. The ingestion of a hearty meal increases the temperature but slightly; an absence of food, as in starvation, produces a marked decrease. Alcoholic drinks, in large amounts, in **persons** unaccustomed to their use, cause a depression of the temperature, amounting from 1° to 2° F. Tea causes a slight elevation.

External temperature. Long continued exposure to cold, especially if the body is at rest, diminishes the temperature from 1° to 2° F., while exposure to a great heat slightly increases it.

Disease frequently causes a marked variation in the normal temperature of the body, rising as high as 107° F. in typhoid fever, and 105° F. in pneumonia; in cholera it falls as low as 80° F. Death usually occurs when the heat remains high and persistent, from 106° to 110° F.; the increase of heat in disease is due to excessive production **rather** than to diminished elimination.

The source of heat is to be sought for in the chemical decompositions and hydrations taking place during the general process of nutrition, and **the** combustion of the carbonaceous compounds by the oxygen of the inspired air; the amount of its production is in proportion to the activity of the internal changes.

Every contraction of a muscle, every act of secretion, each exhibition of nerve force, is accompanied by a change in the chemical composition of the tissues and an evolution of heat. The reduction of the disintegrated tissues to their simplest form by oxidation; the combination of the oxygen of the inspired air with the carbon and hydrogen of the blood and tissues, **results in the** formation of carbonic acid and water and the generation of a **large amount of heat.**

Certain elements of the food, particularly the *non-nitrogenized* substances, **undergo oxidation** without taking part in the formation of the tissues, being **transformed into** carbonic acid and water, and thus increase the sum of heat in the body.

Heat-producing Tissues. All the tissues of the body add to **the** general amount of heat, according to the degree of their activity. But special structures, on account of their mass and the large amount of blood they receive, are particularly to be regarded as *heat producers; e.g.:—*

1. During mental activity the *brain* receives nearly one-fifth of the entire volume of blood, and the venous blood returning from it is charged with waste matters, and its temperature is increased.

2. The *muscular tissue*, on account of the many chemical changes occurring during active contractions, **must** be regarded as the chief heat-producing tissue.

3. The *secreting glands*, during their functional activity, add largely to the amount of heat.

The entire quantity of heat generated within the body has been demonstrated experimentally to be about 2300 calories, a calorie or heat unit being that amount of heat required to raise the temperature of one kilo. of water (2.2 lbs.) one degree Centigrade. This quantity of heat if not utilized and retained within the body would elevate its temperature in 24 hours about 60° F. That this volume of heat depends very largely upon the oxidation of the food stuffs can be shown experimentally.

The normal temperature of the body is maintained by a constant expenditure of the heat in several directions.

1. In warming the food, drink and air that are consumed in 24 hours. For this purpose about 157 heat units are required.

2. In evaporating water from the skin and lungs; 619 heat units being utilized for this purpose.

3. In radiation and conduction. By these processes the body loses at least 50 per cent. of its heat, or 1156 heat units.

4. In the production of work; the work of the circulatory, respiratory, muscular, and nervous apparatus being performed by the transformation of 369 heat units into units of work.

The *nervous system* influences the production of heat in a part, by increasing the amount of blood going through it by its action upon the vasomotor nerves. Whether there exists a special heat centre has not been satisfactorily determined, though this is probable.

SECRETION.

The Process of Secretion consists in the separation of materials from the blood, which are either to be again utilized to fulfill some special purpose in the economy, or are to be removed from the body as excrementitious matter; in the former case they constitute the *secretions*, in the latter, the *excretions*.

The *materials* which enter into the composition of the *secretions* are derived from the nutritive principles of the blood, and require special organs, *e. g.*, gastric glands, mammary glands, etc., for their proper elaboration.

The *materials* which compose the *excretions* preëxist in the blood, and are the results of the activities of the nutritive process; if retained within the body they exert a deleterious influence upon the composition of the blood.

Destruction of a secreting gland abolishes the secretion peculiar to it, and it cannot be formed by any other gland; but among the excreting organs there exists a complementary relation, so that if the function of one organ be interfered with, another performs it, to a certain extent.

CLASSIFICATION OF THE SECRETIONS.

PERMANENT FLUIDS.

Serous fluids.
Synovial fluid.
Aqueous humor of the eye.

Vitreous humor of the eye.
Fluid of the labyrinth of the internal ear.
Cerebro-spinal fluid.

TRANSITORY FLUIDS.

Mucus.
Sebaceous matter.
Cerumen (external meatus).
Meibomian fluid.
Milk and colostrum.
Tears.
Saliva.

Gastric juice.
Pancreatic juice.
Secretion from Brunner's glands.
Secretion from Leiberkühn's glands.
Secretion from follicles of the large intestine.
Bile (also an excretion).

EXCRETIONS.

Perspiration and the secretion of the axillary glands.

Urine.
Bile (also a secretion).

FLUIDS CONTAINING FORMED ANATOMICAL ELEMENTS.

Seminal fluid, containing spermatozoids.   Fluid of the Graafian follicles.

The essential apparatus for secretion is a delicate, homogeneous, structureless *membrane*, on one side of which, in close contact, is a capillary plexus of *blood vessels*, and on the other side a **layer of** *cells* whose physiological function varies in different situations.

Secreting organs may be divided into *membranes* and *glands*.

Serous membranes usually exist as closed sacs, the inner surface of which is covered by pale, nucleated epithelium, containing a small amount of secretion.

The serous membranes are the *pleura, peritoneum, pericardium, synovial sacs*, etc.

The *serous fluids* are of a pale amber color, somewhat viscid, alkaline, coagulable by heat, and resemble the serum of the blood; their *amount* is but small; the pleural varies from 4 to 7 drachms; the peritoneal from 1 to 4 ounces; the pericardial from 1 to 3 drachms.

The *synovial fluid* is colorless, alkaline, and extremely viscid, from the presence of *synovine*.

The *function* of serous fluids is to moisten the opposing surfaces, so as to prevent friction during the play of the viscera.

The *mucous membranes* are soft and velvety in character, and line the cavities and passages leading to the exterior of the body, *e. g.*, the *gastro-intestinal*, *pulmonary* and *genito-urinary*. They consist of a primary basement membrane covered with epithelial cells, which, in some situations, are tessellated, in others, columnar.

Mucus is a pale, semi-transparent, alkaline fluid, containing epithelial cells and leucocytes. It is composed, chemically, of water, an albuminous principle, *mucosine*, and mineral salts; the principal varieties are nasal, bronchial, vaginal and urinary.

Secreting Glands are formed of the same elements as the secreting membranes; but instead of presenting flat surfaces, are involuted, forming tubules, which may be *simple follicles*, *e. g.*, mucous, uterine or intestinal; or *compound follicles*, *e. g.*, gastric glands, mammary glands; or *racemose* glands, *e. g.*, salivary glands and pancreas. They are composed of a basement membrane, enveloped by a plexus of blood vessels, and are lined by epithelial and *true secreting cells*, which in different glands possess the capability of elaborating elements characteristic of their secretions.

In the production of the secretions two essentially different processes are concerned :—

1. *Chemical.* The formation and elaboration of the characteristic organic ingredients of the secreting fluids, *e. g.*, pepsin, pancreatin, takes place during the intervals of glandular activity, as a part of the general function of nutrition. They are formed by the cells lining the glands, and can often be seen in their interior with the aid of the microscope, *e. g.*, bile in the liver cells, fat in the cells of the mammary gland.

2. *Physical.* Consisting of a transudation of water and mineral salts from the blood into the interior of the gland.

During the intervals of glandular activity, only that amount of blood passes through the gland sufficient for proper nutrition; when the gland begins to secrete, under the influence of an appropriate stimulus, the blood vessels dilate and the quantity of blood becomes greatly increased beyond that flowing through the gland during its repose.

Under these conditions a transudation of water and salts takes place, washing out the characteristic ingredients, which are discharged by the gland ducts. The *discharge* of the secretions is intermittent; they are retained in the glands until they receive the appropriate stimulus, when they pass into the larger ducts by the vis-a-tergo, and are then discharged by the contraction of the muscular walls of the ducts.

The *activity* of glandular secretion is hastened by an increase in the blood pressure and retarded by a diminution.

The *nervous centres* in the medulla oblongata influence secretion, (1) by increasing or diminishing the amount of blood entering a gland; (2) by exerting a direct influence upon the secreting cells themselves, the centres being excited by reflex irritation, mental emotion, etc.

MAMMARY GLANDS.

The Mammary Glands secrete the milk, and undergo at different periods of life remarkable changes in structure. Though rudimentary in childhood, they gradually increase in size as the young female approaches puberty.

The gland presents, at its convexity, a small prominence of skin, the *nipple*, surrounded by an *areola* of a deeper tint. It is covered anteriorly by a layer of adipose tissue and posteriorly by a fibrous structure which attaches it loosely to the *pectoralis* muscle.

During *utero-gestation* the mammæ become large, firm, well-developed and lobulated; the areola becomes darker and the veins more prominent. In the intervals of lactation the glands gradually shrink in size to their original condition, undergo involution, and become non-secreting organs.

Structure of the Mammæ. The mamma is a conglomerate gland, consisting of a number of *lobes*, from 15 to 20 in number, each of which is subdivided into *lobules* made up of *gland vesicles* or *acini*. The ducts which convey the secretion to the exterior, the *lactiferous ducts*, open by 15 to 20 orifices upon the surface of the nipple, at the base of which they are dilated to form little reservoirs in which the milk collects during the periods of active secretion.

The *walls* of the lacteal duct consist of white, fibrous tissue, and non-striated muscular fibres, lined by short columnar cells, which disappear during active lactation. The ducts measure about the $\frac{1}{12}$ of an inch in diameter; as they pass into the substance of the gland, each duct divides into a number of branches, which are distributed to distinct lobules and terminate in the acini.

An *acinus* is made up of a number of *vesicles* composed of a homogeneous membrane, lined by pavement epithelium. The gland vesicles are held together by white, fibrous tissue, which unites the lobules into lobes.

MILK.

Milk has a pale blue color, is almost inodorous, of a sweetish taste, an alkaline reaction, and a specific gravity varying from 1.025 to 1.046. Examined microscopically it is seen to contain an immense number of globules, measuring the $\frac{1}{10500}$ of an inch in diameter, suspended in a clear fluid; these are the *milk globules*, formed of a small **mass of** oily matter covered by a layer of albumen.

The *quantity* of milk secreted by the human female in 24 hours, during **the period of** lactation, is about two to three pints; the quantity removed **by the infant** from a full breast at one time being about two ounces.

COMPOSITION OF MILK.

Water	890.00
Proteids, including casein and serum albumen	35.00
Fatty matter (butter)	25.00
Sugar (lactose) with extractives	48.00
Salts	2.00
	1000.00

Casein is the nutritive principle of milk, and constitutes its most important ingredient. It is held in solution by an alkali, but upon the addition of an acid it undergoes coagulation, passing into a semi-solid form. The presence of lactic acid, resulting from a transformation of milk sugar, causes spontaneous coagulation to take place.

The *Fatty matter* is more or less solid at ordinary temperature, and consists of margarine and oleine; when subjected to the churning process the globules run together and form a coherent mass, the *butter*.

When milk is allowed to stand for a varying length of time the fat globules rise to the surface, forming a layer more or less thick, the *cream*.

Milk sugar or lactose is an important ingredient in the food of the young child; it is readily transformed into lactic acid in the presence of nitrogenized **ferments**.

Influences modifying the secretion. During lactation there is a demand for an increased amount of fluid, and if not supplied, the amount of milk secreted is diminished. Good food in sufficient quantity is necessary for the proper elaboration of milk, though no particular article influences its production.

Mental emotion at times influences the character of the milk, decreasing the amount of its different constituents.

Mechanism of Secretion. The water and salts preëxist in the blood **and pass into the** gland vesicles by *osmosis*. The casein, fatty matter and sugar appear only in the mammary gland, but the mechanism of their formation is not understood.

Colostrum is a yellowish, opaque fluid, formed in the mammary glands toward the latter period of utero-gestation; it consists of water, albumen, fat, sugar and salts, and acts as a laxative to the newly-born infant.

VASCULAR OR DUCTLESS GLANDS.

The Vascular Glands are regarded as possessing the power of acting upon certain elements of the food and aiding the process of sanguinification; of modifying the composition of the blood as it flows through their substance, by some act of secretion.

The vascular glands are the *spleen, supra-renal capsules,* **thyroid** and *thymus* glands.

The Spleen is about 5 inches in length, 6 ounces in weight, of a dark bluish color, and situated in the left hypochondriac region. It is covered externally by a reflection of the peritoneum, beneath which is the proper fibrous coat, composed of areolar and elastic tissue and non-striated muscular fibres. From the inner surface of the fibrous envelope processes or trabeculæ are given off, which penetrate the substance of the gland, forming a network, in the meshes of which is contained the *spleen pulp*. The *splenic* **artery** divides into a number of branches, some of which, when they become very minute, pass directly into veins, while others terminate in true capillaries.

As the capillary vessels ramify through the substance of the gland, their walls frequently disappear and the blood passes from the arteries into the veins through *lacunæ* (Gray).

The *splenic* or *Malpighian corpuscles* are small bodies, spherical or ovoid in shape, the $\frac{1}{75}$ of an inch in diameter, situated upon the sheaths of the small arteries. They consist of a delicate membrane, containing a semi-fluid substance composed of numerous small cells resembling lymph corpuscles. The *spleen pulp* is a dark red, semi-fluid substance, of a soft consistence, contained in the meshes of the trabeculæ. In it are found numerous corpuscles, like those observed in the Malpighian bodies, blood corpuscles in a natural and altered condition, nuclei and pigment granules.

Function of the Spleen. Probably influences the preparation of the albuminous food for nutrition; during digestion the spleen becomes larger, its contents are increased in amount, and after digestion it gradually diminishes in size, returning to the normal condition.

The *red corpuscles* are here disintegrated, after having fulfilled their function in the blood; the splenic venous blood containing relatively a small quantity.

The *white corpuscles* appear to be increased in number, the blood of the splenic vein containing an unusually large proportion.

The *spleen* serves also as a reservoir for blood when the portal circulation becomes obstructed.

The *nervous system* controls the enlargement of the spleen; division of the nerve produces dilatation of the vessels, stimulation contracts them.

The Supra-renal Capsules are triangular, flattened bodies, situated above the kidney. They are invested by a fibrous capsule sending in trabeculæ, forming the framework. The glandular tissue is composed of two portions, a *cortical* and *medullary*. The cortical being made up of small cylinders lined by cells and containing an opaque mass, nuclei and granular matter. The medullary consists of a fibrous network containing in the alveoli nucleated protoplasm.

The Thyroid gland consists of a fibrous stroma, containing ovoid closed sacs, measuring on the average $\frac{1}{40}$ of an inch, formed of a delicate membrane lined by cells; the contents of the sacs consist of yellowish albuminous fluid.

The Thymus gland is most developed in early life and almost disappears in the adult. It is divided by processes of fibrous tissue into lobules, and these again into follicles which contain lymphoid corpuscles.

The *functions* of the vascular glands appear to be the more complete elaboration of the blood necessary for proper nutrition; they are most highly developed during infancy and embryonic life, when growth and development are most active.

EXCRETION.

The Principal Excrementitious Fluids discharged from the body are the urine, perspiration and bile; they hold in solution principles of waste which are generated during the activity of the nutritive process, and are the ultimate forms to which the organic constituents are reduced in the body. They also contain inorganic salts.

The Urinary Apparatus consists of the kidneys, ureters and bladder.

KIDNEYS.

The **Kidneys** are the organs for the secretion of urine ; they resemble a bean in shape, are from four to five inches in length, two in breadth, and weigh from four to six ounces.

FIG. 8.

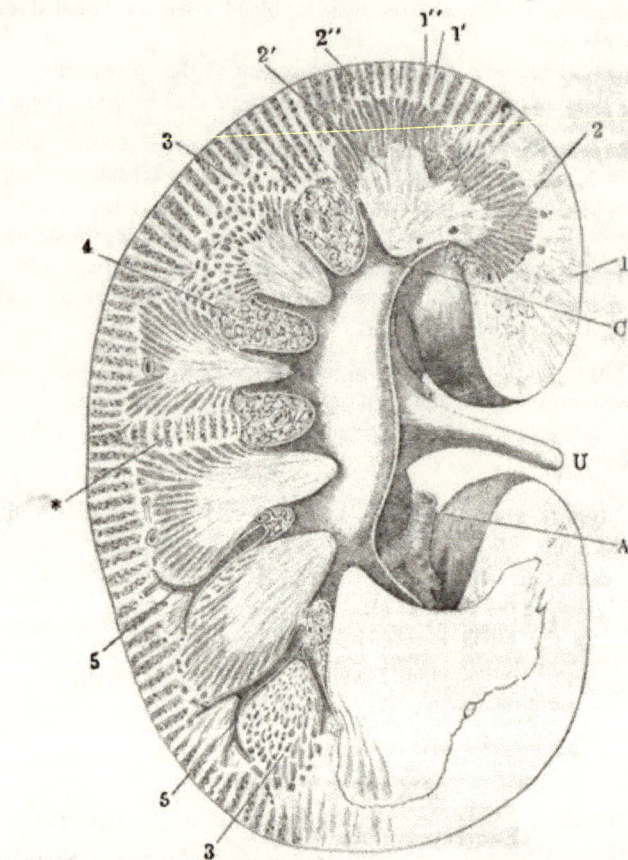

Longitudinal section through the kidney, the pelvis of the kidney, and a number of renal calyces. A, branch of the renal artery ; **U**, ureter ; **C**, renal calyx ; 1, cortex ; 1', medullary rays ; 1", labyrinth, or cortex proper ; **2**, medulla ; 2', papillary portion of medulla, or medulla proper ; 2", border layer of the medulla ; 3, 3, transverse section through the axes of the tubules of the border layer ; 4, fat of the renal sinus . 5, 5, arterial branches ; *, transversely coursing medulla rays.—*Tyson, after Henle.*

They are situated in the lumbar region, one on each side of the vertebral column, behind the peritoneum, and extend from the 11th rib to the crest of the ilium; the anterior surface is convex, the posterior concave, and presents a deep notch, the *hilum*.

The kidney is surrounded by a thick layer of fat, beneath which is the fibrous coat, thin and smooth, composed of dense white fibrous tissue with which are intermingled elastic fibres. It is adherent to the surface of the organ, but can easily be removed by dissection.

The Substance of the Kidney is dense, but friable; upon making a longitudinal section, and dividing it, there is presented a cavity, the *pelvis*, lined by the proper fibrous coat and occupied by the expanded portion of the *ureter*.

The kidney exhibits two structures, viz. :—

1. *An external* or *cortical portion*, about ⅙ of an inch in diameter, of a reddish color, and somewhat granular.

2. *An internal* or *medullary* portion, of a dark red color, arranged in the form of *pyramids*, the bases of which are directed toward the cortical portion, and the apices toward the pelvis, into which they project, and are covered by the calyces.

The Cortical portion of the kidney consists of a delicate matrix containing an immense number of tubules, having a markedly convoluted appearance, and interlacing in every direction (the tubules of Ferrein). Throughout its structure are numerous ovoid bodies, *the Malpighian bodies*, which are the flask-like terminations of the convoluted tubules; these tubes are composed of a delicate homogeneous membrane lined by nucleated cells. After pursuing a most intricate course in the cortical portion, they become narrower and form loops which dip into the pyramidal portion (Henle's tubules), returning upon themselves, to finally terminate in the straight tubes of the pyramids.

The Malpighian bodies, the dilated extremities of the convoluted tubes, consist of a little sac (the capsule of Müller), which is ovoid in shape, measuring about the 1/100 of an inch in diameter, and contains a tufted mass of minute blood vessels, over the surface of which is reflected a layer of cells.

Medullary Substance. The conical masses, the *pyramids of Mal-*

FIG. 9.

Diagrammatic exposition of the method in which the uriniferous tubes unite to form primitive cones. — *Tyson, after Ludwig.*

pighi, consist of a number of straight tubes, which commence at the apex by from 10 to 20 openings; and **as they pass toward** the cortical portion, they divide and subdivide at acute angles, until a **large mass** of tubes is produced. These tubes are on the average about $\frac{1}{500}$ of an inch in diameter, and composed of a thin, but firm, elastic, structureless membrane, lined by polygonal nucleated cells, which reduce the diameter of the **lumen** of the tube about two-thirds; these are the *straight tubes* of Bellini.

Blood vessels of the Kidney. *The renal artery* is of large size and enters the organ at the hilum; it divides into several large branches, which **penetrate** the substance of the kidney, between the pyramids, at the base **of** which they **form** an anastomosing plexus, which completely surrounds **them.** From this plexus vessels follow **the** straight tubes toward the apex, **while others,** entering the cortical portion, divide into small **twigs** which **enter** the Malpighian body and form **a** mass of convoluted vessels, the *glomerulus*. After circulating through the Malpighian tuft the blood **is** gathered together by two or three small veins, which again subdivide and form a fine capillary plexus, which envelops the convoluted tubules; from this plexus the veins converge to form the emulgent vein, which empties **into** the vena cava.

The nerves of the kidney follow **the course of the blood vessels and** are derived from the renal plexus.

The Ureter is a membranous tube, situated behind the peritoneum, about the diameter of a goose quill, 18 inches in length, and extends from **the** pelvis of the kidney to the base **of** the bladder, which it perforates in an oblique direction. It is composed of 3 coats, fibrous, muscular **and** mucous.

The Bladder is a reservoir **for** the temporary reception of the urine **prior to** its expulsion from the body; when fully distended it is ovoid in **shape,** and holds about *one pint*. It is composed of four coats, *serous, muscular,* the fibres of which are arranged longitudinally and circularly, *areolar* and *mucous*. The orifice of the bladder is controlled by the *sphincter vesicæ,* a muscular band about half an inch in width.

As soon as the urine is formed it passes through the *tubuli uriniferi* into the pelvis, and from thence through the ureters into the bladder, which it enters at an irregular rate. Shortly after a meal, after the ingestion of large quantities of fluid, and after exercise, the urine flows into the blad-der quite rapidly, while it is reduced to a few drops during the intervals of digestion. It is prevented from regurgitating into the ureters on account of the oblique direction they take between the mucous and muscular coats.

Nervous Mechanism of Urination. When the urine has passed into the bladder, it is there retained by the sphincter vesicæ muscle, kept **in a** state of tonic contraction by the action of a nerve centre in the lumbar region of the spinal cord. This centre can be inhibited and the sphincter relaxed, either *reflexly*, by impressions coming through sensory nerves from the mucous membrane of the bladder, or *directly*, by a voluntary impulse descending the spinal cord. When the desire to urinate is experienced, impressions made upon the vesical sensory nerves are carried to the centres governing the *sphincter* and *detrusor urinæ* muscles and to the brain. If now the act of urination is to take place, a voluntary impulse, originating in the brain, passes down the spinal cord and still further inhibits the sphincter vesicæ centre, with the effect of relaxing the muscle, and of stimulating the centre governing the detrusor muscle, with the effect of contracting the muscle and expelling the urine. If the act is to be suppressed, voluntary impulses inhibit the *detrusor* centre and possibly stimulate the *sphincter* centre.

The *genito-spinal* centre controlling these movements is situated in that portion of the spinal cord corresponding to the origin of the 3d, 4th and 5th sacral nerves.

URINE.

Normal Urine is of a pale yellow or amber color, perfectly transparent, with an aromatic odor, an acid reaction, a specific gravity of 1.020, and a temperature when first discharged of 100° Fahr.

The *color* varies considerably in health, from a pale yellow to a brown hue, due to the presence of the coloring matter, *urobilin* or *urochrome*.

The *transparency* is diminished by the presence of mucus, the calcium and magnesium phosphates and the mixed urates.

The *reaction* is slightly acid, caused by the acid phosphate of sodium. After standing for a short time, an increased acidity is observed, due to an acid fermentation, from the presence of mucus. The urea is converted into ammonium carbonate, giving rise to a strong ammoniacal odor.

The *specific gravity* varies from 1.010 to 1.025.

The *quantity* of urine excreted in 24 hours is between 40 and 50 fluid ounces, but ranges above and below this standard.

The *odor* is characteristic, and caused by the presence of taurylic and phenylic acids, but is influenced by vegetable foods and other substances eliminated by the kidneys.

COMPOSITION OF URINE.

Water... 967.

Urea.. 14.230

Other nitrogenized crystalline bodies, uric acid, principally in the form of alkaline urates.

Creatin, creatinin, xanthin, hypoxanthin.

Hippuric acid, leucin, tyrosin, taurin, cystin, all in small amounts, and not constant. } 10.635

Mucus and pigment.

Salts :—

Inorganic, principally sodium and potassium sulphates, phosphates and chlorides, with magnesium and calcium phosphates, traces of silicates and chlorides.

Organic : lactates, hippurates, acetates, formates, which appear only occasionally. } 8.135

Sugar.. a trace.

Gases (nitrogen and carbonic acid principally).

_____

1000.00

The **Average Quantity** of the principal constituents excreted in 24 hours is as follows :—

Water... 52 fluid oz.

Urea.. 512.4 grains.

Uric acid... 8.5 "

Phosphoric acid... 45.0 "

Sulphuric acid.. 31.11 "

Inorganic salts... 323.25 "

Lime and magnesia...................................... 6.5 "

To **Determine the amount** of solid matters in any given amount of urine, multiply the last two figures of the specific gravity by the coefficient of Hæser, 2.33; _e. g._, in 1000 grains of urine having a specific gravity 1.022, there are contained 22 × 2.33 = 51.26 grains of solid matter.

The **Elimination** of the urinary constituents is accomplished by the two processes of filtration and secretion.

1. **By Filtration** the water and mineral salts are removed from the blood; this takes place, for the most part, in the Malpighian corpuscles, by the process of osmosis. The _amount_ of these constituents eliminated varies with the pressure of blood in the renal arteries. All of the agencies which increase the general blood pressure increase the quantity of urine.

Season. In summer, while the capillary vessels of the skin are dilated, and perspiration is abundant, there is a diminished blood pressure, and a consequent diminution in the amount of urine; in winter the reverse takes place.

During sleep the renal excretion is diminished, but increased in the morning hours, and especially after the ingestion of hearty meals.

The nervous system influences the secretion of urine. Irritation of the medulla oblongata, a little above the origin of the pneumogastric and auditory nerves, increases the quantity; division of the renal nerves destroys the nutrition of the kidney, and thus interferes with the elimination of the urine. Mental emotion, fear, anxiety, etc., increase the amount secreted.

2. Secretion. While it is established that the Malpighian corpuscles permit the filtration of water and salts, it has also been shown that the renal epithelial cells lining the convoluted tubes are the agencies by which the solid matters, *urea*, *creatin*, etc., are removed from the blood, by a process of *true secretion*, which is independent of blood pressure and caused by the presence of these ingredients in the blood.

Urea is the most important of the organic constituents of the urine. It is a colorless, neutral substance, crystallizing in four-sided prisms, soluble in boiling alcohol and water; when subjected to prolonged boiling it is decomposed, with the production of ammonium carbonate.

Urea is not formed in the kidneys, but pre-exists in the blood.

The Amount of Urea excreted in 24 hours is estimated at about 500 grains; it is *increased* during the waking hours, by an animal diet and by prolonged muscular exertion ; *diminished* during sleep and by non-nitrogenized food.

Source. Urea results from an imperfect oxidation of the albuminous principles of the food, and from a disintegration of the organic constituents of the tissues.

Uric acid, or *lithic acid*, is a constant ingredient of the urine; the amount excreted daily is about 8 grains; it is increased by nitrogenized, decreased by non-nitrogenized food. It exists in the urine in a free state, and as the urate of soda. It arises from the disassimilation of albuminous compounds, and when secreted in excess is deposited in a crystalline form, as a brown or "brick-red" sediment, with the sodium and ammonium urates.

Creatin is a colorless, transparent substance, crystallizing in prisms ; found in blood, kidneys, and muscular tissue; by boiling in acid solutions it is transformed into

Creatinin, which resembles creatin chemically. It is soluble in water and alcohol, and crystallizes in colorless prisms. About 15 grains are excreted daily.

The *Earthy phosphates* are insoluble in water but held in solution in the urine by the acid reaction. If the **urine becomes** alkaline, they are deposited copiously, and yet may not be increased in quantity; from 15 to 25 grains are excreted in 24 hours. The *sulphates* are those of sodium and potassium; they are very soluble and do not appear as a precipitate; the average quantity excreted in 24 hours is about 60 grains.

Abnormal ingredients appear in the urine at times, in pathological conditions, *e. g.*, sugar, albumen, biliary salts, etc.

The *Gases* of the urine are carbonic acid and nitrogen.

LIVER.

The Liver is a highly vascular, conglomerate gland, appended to the alimentary canal, and performs the triple office of (1) *excreting bile*, (2) *elaborating blood* and (3) *secreting glycogen.*

It is the largest gland in the body, weighing about 4½ pounds; it **is** situated in the right hypochondriac region, and retained in position by five ligaments, four of which are formed by duplicatures of the peritoneal investment.

The *proper coat* of the liver is a thin but firm fibrous membrane, closely adherent to the surface of the organ, which it penetrates at the transverse fissure, and follows the vessels in their ramifications through its substance, constituting *Glisson's capsule.*

Structure of the Liver. The liver is made up of a large **number of** small bodies, *the lobules*, rounded or ovoid in shape, measuring the $\frac{1}{15}$ of an inch in diameter, separated by a space in which are situated blood vessels, nerves, hepatic ducts and lymphatics.

The **lobules** are composed of cells, which, when examined microscopically, exhibit a rounded or polygonal shape, and measure, on the average, the $\frac{1}{1000}$ of an inch in diameter; they possess one, and at times two, nuclei; they also contain globules of fat, pigment matter, and animal starch. The cells constitute the *secreting* structure of the liver, and are the *true hepatic cells.*

The Blood vessels which enter the liver **are** (1) The *portal vein*, made up of the *gastric, splenic, superior* and *inferior mesenteric veins*; (2) the *hepatic artery*, a branch of the cœliac axis; both of which are invested by a sheath of areolar tissue; the vessels which leave the liver are the *hepatic veins*, originating in its interior, collecting the blood distributed by the portal vein and hepatic artery, and conducting it to the *ascending vena cava.*

Distribution of Vessels. The *portal vein* and *hepatic artery*, upon entering the liver, penetrate its substance, divide into smaller and smaller branches, occupy the spaces between the lobules, completely surrounding and limiting them, and constitute the *interlobular* vessels. The *hepatic artery*, in its course, gives off branches to the walls of the portal vein and Glisson's capsule, and finally empties into the small branches of the *portal vein* in the interlobular spaces.

The *interlobular vessels* form a rich plexus around the lobules, from which branches pass to neighboring lobules and enter their substance, where they form a very fine network of capillary vessels, ramifying over the hepatic cells, in which the various functions of the liver are performed. The blood is then collected by small veins, converging toward the centre of the lobule, to form the *intralobular vein*, which runs through its long axis and empties into the *sub-lobular vein*. The *hepatic veins* are formed by the union of the sub-lobular veins, and carry the blood to the ascending vena cava; their walls are thin and adherent to the substance of the hepatic tissue.

The **Hepatic Ducts or Bile Capillaries** originate within the lobules, in a very fine plexus lying between the hepatic cells; whether the smallest vessels have distinct membranous walls, or whether they originate in the spaces between the cells by open orifices, has not been satisfactorily determined.

The *Bile Channels* empty into the interlobular ducts, which measure about $\frac{1}{2000}$ of an inch in diameter, and are composed of a thin homogeneous membrane lined by flattened epithelial cells.

As the interlobular bile ducts unite to form larger trunks, they receive an external coat of fibrous tissue, which strengthens their walls; they finally unite to form one large duct, the *hepatic duct*, which joins the *cystic duct ;* the union of the two forms the *ductus communis choledochus*, which is about three inches in length, the size of a goose quill, and opens into the duodenum.

The **Gall Bladder is a** pear-shaped sack, about four inches in length, situated in a fossa on the under surface of the liver. It is a reservoir for the bile, and is capable of holding about one ounce and a half of fluid. It is composed of three coats, (1) serous, a reflection of the peritoneum, (2) fibrous and muscular, (3) mucous.

(1) Bile. **Mechanism of its Secretion.** Bile does not preëxist in the blood, but is formed in the interior of the hepatic cells, from materials derived from the venous as well as arterial blood. The secreted bile is

then taken up by the delicate plexus of ve-sels, from which it passes into the larger ducts, and finally either empties into the intestine or is regurgitated backward into the gall bladder, in which it is stored up during the intervals of digestion.

Although the secretion of bile is constantly taking place, it is only when the food passes into the intestinal canal that this fluid is discharged abundantly, under the influence of the contraction of the walls of the gall bladder; it increases in amount during the period of active digestion, from the second to the eighth hour, and then gradually diminishes.

The Bile is both a *secretion* and an *excretion;* it contains new constituents which are formed only in the substance of the liver, and are destined to play an important part ultimately in nutrition; it contains also waste ingredients which are discharged into the intestinal canal and eliminated from the body.

The physical properties and functions of bile have been considered under the head of digestion (see page 34).

(2) **Elaboration of Blood.** Besides the capability of secreting bile, the liver possesses the property of so acting upon and modifying the chemical composition of the products of digestion, as they traverse its substance, that they readily assimilate with the blood, and are transformed into materials capable of being converted into the elements of the blood and solid tissues.

The *albuminose* particularly requires the modifying influence of the liver; for if it be removed from the portal vein and introduced into the jugular vein, it is at once removed from the blood by the action of the kidneys.

The blood of the *hepatic vein* differs from the blood of the *portal vein*, in being richer in blood corpuscles, both red and white; its plasma is more dense, containing a less percentage of water and a greater amount of solid constituents, but no fibrin; its serum contains less albumen, fat and salts, but its sugar is increased.

(3) **Glycogenic Function.** In addition to the two preceding functions, Bernard, in 1848, demonstrated the fact that the liver, during life, normally produces a sugar-forming substance, analogous in its chemical composition to starch, which he termed glycogen; also that when the liver is removed from the body, and its blood vessels thoroughly washed out, after a few hours sugar again makes its appearance, in abundance.

It can be shown to exist in the blood of the hepatic vein as well as in a decoction of the liver substance, by means of either Trommer's or Fehling's tests, even when the blood of the portal vein does not contain a trace of sugar.

Origin and Destination of Glycogen. Glycogen appears to be formed *de novo* in the liver cells, from materials derived from the food, whether the diet be animal or vegetable, though a larger per cent. is formed when the animal is fed on starchy and saccharine, than when fed on animal food. The *glucose*, which is one of the products of digestion, is absorbed by the blood vessels, and carried directly into the liver; as it does not appear in the urine, as it would if injected at once into the general circulation, it is probable that it is detained in the liver, dehydrated and stored up as *glycogen*. The change is shown by the following formula :—

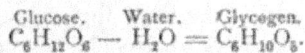

$$\underset{\text{Glucose.}}{C_6H_{12}O_6} - \underset{\text{Water.}}{H_2O} = \underset{\text{Glycogen.}}{C_6H_{10}O_5}.$$

The glycogen thus formed is stored up in the hepatic cells for the future requirements of the system. When it is carried from the liver it is again transformed into *glucose* by the agency of a ferment. Glycogen does not undergo oxidation in the blood ; this takes place in the tissues, particularly in the muscles, where it generates heat and contributes to the development of muscular force.

Glycogen, when obtained from the liver, is an amorphous, starch-like substance, of a white color, tasteless and odorless, and soluble in water; by boiling with dilute acids, or subjected to the action of an animal ferment, it is easily converted into *glucose*. When an excess of sugar is generated by the liver, it can be found, not only in the blood of the hepatic vein, but also in other portions of the body; under these circumstances it is eliminated by the kidneys, appearing in the urine, constituting the condition of *glycosuria*.

The Nervous System influences the production of the glycogenic matter ; irritation of the medulla oblongata, between the auditory and pneumogastric nerves, is followed by an increase in the production of sugar, and its appearance in the urine, which, however, is only temporary.

SKIN.

The Skin, the external investment of the body, is a most complex and important structure, serving (1) as a *protective covering ;* (2) an organ for *tactile sensibility ;* (3) an organ for the *elimination of excrementitious matters.*

The Amount of Skin investing the body of a man of average size is about twenty feet, and varies in thickness, in different situations, from the $\frac{1}{8}$ to the $\frac{1}{100}$ of an inch.

The skin consists of two principal layers, viz., a deeper portion, the *Corium*, and a superficial portion, the *Epidermis*.

The Corium, or Cutis Vera, may be subdivided into a *reticulated* and a *papillary* layer. The *former* is composed of white fibrous tissue, non-striated muscular fibres and elastic tissue, interwoven in every direction, forming an areolar network, in the meshes of which are deposited masses of fat, and a structureless amorphous matter; the *latter* is formed mainly of club-shaped elevations or projections of the amorphous matter, constituting the *papillæ;* they are most abundant, and well developed, upon the palms of the hands and the soles of the feet; they average the $\frac{1}{100}$ of an inch in length, and may be simple or compound; they are well supplied with nerves, blood vessels and lymphatics.

The Epidermis or Scarf Skin is an extra-vascular structure, a product of the true skin, and composed of several layers of cells. It may be divided into two layers, the *rete mucosum* or the *Malpighian layer*, and the *horny* or *corneous*.

The *former* closely applies itself to the papillary layer of the true skin, and is composed of large, nucleated cells, the lowest layer of which, the "prickle cells," contain pigment granules, which give to the skin its varying tints in different individuals and in different **races** of men; the more superficial cells are large, colorless, and semi-transparent. The *latter*, the corneous layer, is composed of flattened cells, which, from their exposure to the atmosphere, are hard and horny in texture; it varies in thickness from $\frac{1}{8}$ of an inch on the palms of the hands and feet, to the $\frac{1}{500}$ of **an inch** in the external auditory canal.

APPENDAGES OF THE SKIN.

Hairs are found in almost all portions of the body, and can be divided into (1) long, soft hairs, on the head; (2) short, stiff hairs, along the edges of the eyelids and nostrils; (3) soft, downy hairs, on the general cutaneous **surface.** They consist of a *root* and a *shaft*, which is oval in shape, and about the $\frac{1}{400}$ of an inch in diameter; it consists of fibrous tissue, covered externally by a layer of imbricated cells, and internally by cells containing granular and pigment material.

The *Root* of the hair is embedded in the hair follicle, formed by a tubular depression of the skin, extending nearly through to the subcutaneous tissue; its walls are formed by the layers of the corium, covered by epidermic cells.

At the bottom of the follicle is a papillary projection of amorphous matter, corresponding to a papilla of the true skin, containing blood vessels and nerves, upon which the hair root rests. The investments of the hair roots are formed of epithelial cells, constituting the *internal* and *external* root sheaths.

The hair protects the head from the heat of the sun **and** cold, retains the heat of the body, prevents the entrance of foreign matter into the lungs, nose, ears, etc. The *color* is due to the pigment matter, which, in old age, becomes more or less whitened.

The Sebaceous Glands, imbedded **in the true** skin, are **simple and** compound racemose glands, opening, **by** a common excretory duct, upon **the** surface of the epidermis or into the hair follicle. They are found **in all** portions of the body, most abundantly in the face, and are formed by a delicate, structureless membrane, lined by flattened polyhedral cells. The sebaceous glands secrete a peculiar oily matter, the *sebum*, by which the skin is lubricated and the hairs softened; it is quite abundant in the region of the nose and forehead, which often present a greasy, glistening appearance; it consists of water, mineral salts, fatty globules, and epithelial cells.

The *Vernix caseosa* which frequently covers the surface of the fœtus at birth consists of the residue of the sebaceous matters, containing epithelial cells and fatty matters; it seems to keep the skin soft and supple, and guards it from the effects of the long continued action of water.

The Sudoriparous Glands excrete the sweat ; they consist of a mass or coil of a tubular gland **duct, situated in the derma and in the subcutane-ous tissue;** average the $\frac{1}{75}$ **of an inch in diameter, and are surrounded by a rich plexus** of capillary **blood vessels.** From this coil the duct **passes in a** straight **direction up through the** skin **to the epidermis, where it makes a few spiral turns and** opens obliquely upon **the surface. The sweat** glands consist **of a** delicate homogeneous membrane lined by epithelial cells, whose function is to extract from the blood the elements existing in the perspiration.

The glands are very abundant all over **the** cutaneous surface, as many as 3528 to the square inch, according to Erasmus Wilson.

The Perspiration is an excrementitious fluid, clear, colorless, almost odorless, slightly acid in reaction, with **a specific** gravity of 1.003 or 1.004.

The total quantity of perspiration excreted daily has been estimated **at about** two pounds, though the amount varies with the nature of the food **and drink,** exercise, external temperature, season, etc.

The elimination of the sweat is not intermittent, **but** continuous ; but it takes place so gradually that as fast as it is formed it passes off by evaporation as *insensible* perspiration.   Under exposure to great heat and exercise the evaporation is not sufficiently rapid, and it appears as *sensible* perspiration.

COMPOSITION OF SWEAT.

Water	995.573
Urea	0.043
Fatty matters	0.014
Alkaline lactates	0.317
Alkaline sudorates	1.562
Inorganic salts	2.491
	1000.00

Urea is a constant ingredient.

Carbonic acid is also exhaled from the skin, the amount being about $\frac{1}{200}$ of that from the lungs.

Perspiration regulates the temperature, and removes waste matters from the blood; it is so important, that if elimination be prevented death occurs in a short time.

The nervous system influences the secretion of watery vapor by causing a dilatation of the capillary blood vessels around the tubular coil.   It is increased by mental emotions; section of the sympathetic fibres in the neck is followed by a copious perspiration; stimulation of the nerves producing contraction of the vessels is followed by an arrestation of the elimination of the sweat.

NERVOUS SYSTEM.

The Nervous System coördinates all the various organs and tissues of the body, and brings the individual into conscious relationship with external nature by means of sensation, motion, language, mental and moral manifestations.

The Nervous Tissue may be divided into two systems, the *Cerebro-spinal* and the *Sympathetic*.

(1) The Cerebro-spinal System occupies the cavities of the cranium and spinal canal, and consists of the brain, the spinal cord, the cranial and spinal nerves. It is the system of animal life, and presides over the functions of sensation, motion, etc.

(2) The Sympathetic System, situated along each side of the spinal column, consists (1) of a double chain of ganglia, united together by nerve cords, and extends from the base of the cranium to the coccyx ; (2) of various ganglia, situated in the head and face, thorax, abdomen, pelvis, etc. All the ganglia are united together by numerous communicating fibres, many of which anastomose with the fibres of the cerebro-spinal system. It is the nervous system of organic life, and governs the functions of nutrition, growth, etc.

Nervous Tissue is composed of two kinds of matter, the *gray* and *white*, which differ in their color, structure and physiological endowments ; the former consists of *vesicles* or *cells* which receive and generate nerve force ; the latter consists of *fibres* which simply conduct it, either from the periphery to the centre or the reverse.

Structure of Gray Matter. The gray matter, found on the surface of the brain in the convolutions, in the interior of the spinal cord, and in the various ganglia of the cerebro-spinal and sympathetic nervous systems, consists of a fine connective tissue stroma, the *neuroglia*, in the meshes of which are embedded the *gray cells* or *vesicles*.

The *cells* are grayish in color, and consist of a delicate investing capsule containing a soft, granular, albuminous matter, a *nucleus*, and sometimes a *nucleolus*. Some of the cells are spherical or oval in shape, while others have an interrupted outline, on account of having one, two or more processes issuing from them, constituting the *uni-polar, bi-polar* or *multi-polar* nerve cells. Cells vary in size ; the smallest being found in the brain, the largest in the anterior horns of gray matter of the spinal cord. Some of the cell processes become continuous with the fibres of the white matter, while others anastomose with those of adjoining cells and form a *plexus*.

Structure of the White Matter. The white matter, found for the most part in the interior of the brain, on the surface of the spinal cord, and in almost all the nerves of the cerebro-spinal and sympathetic systems, consists of minute tubules or fibres, the *ultimate nerve filaments*, which in the perfectly fresh condition, are apparently structureless and homogeneous; but when carefully examined after death are seen to consist of three distinct portions, (1) a tubular membrane; (2) the white substance of Schwann; (3) the axis cylinder.

The *Tubular membrane*, investing the nerve filament, is thin, homogeneous, and lined by large, oval nuclei, and presents, in its course, annular constrictions; it serves to keep the internal parts of the fibre in position, and protects them from injury.

The *White substance* of Schwann, or the *medullary layer*, is situated immediately within the tubular membrane, and gives to the nerves their peculiar white and glistening appearance. It is composed of oleaginous matter in a more or less fluid condition; after death it undergoes coagulation, giving to the fibre a knotted or varicose appearance. It serves to insulate the axis cylinder, and prevents the diffusion of the nerve force.

The *Axis cylinder* occupies the centre of the medullary substance. In the natural condition it is transparent and invisible, but when treated with proper reagents, it presents itself as a pale, granular, flattened band, albuminous in character, more or less solid, and somewhat elastic. It is composed of a number of minute fibrillæ united together to form a single bundle. (Schultze.)

Nerve fibres in which these three structural elements coexist are known as the *medullated* nerve fibres. In the sympathetic system, and in the gray substance of the cerebro-spinal system, many nerves are destitute of a medullary layer, and are known as the *non-medullated* nerve fibres.

Gray or Gelatinous nerve fibres, found principally in the sympathetic system, are gray in color, semi-transparent, flattened, with distinct borders, finely granular, and present oval nuclei.

The diameter of the gelatinous fibres is about the $\frac{1}{8000}$ of an inch; of the medullated fibres from $\frac{1}{1500}$ to $\frac{1}{1200}$ of an inch.

Ganglia are small bodies, varying considerably in size, situated on the posterior roots of spinal nerves, on the sensory cranial nerves, alongside of the vertebral column, forming a connecting chain, and in the different viscera. They consist of a dense, investing, fibrous membrane, containing in its interior *gray* or *vesicular* cells, among which are found *white* and *gelatinous* nerve fibres. They may be regarded as independent nerve centres.

Structures of Nerves. Nerves are rounded or flattened cords extending from the centres to the periphery; they are surrounded externally by a sheath, the *neurilemma*, composed of fibrous and elastic tissue forming a stroma, in which blood vessels ramify, from which the nerves derive their nourishment.

A *Nerve* consists of a greater or less number of *ultimate nerve filaments*, separated into bundles by fibrous septa given off from the neurilemma. The nerve filaments pursue an uninterrupted course, from their origin to their termination; branches pass from one nerve trunk into the sheath of another, but there is no anastomosis or coalescence with adjoining nerve fibres.

A **Plexus** is formed by a number of branches of different nerves interlacing in every direction, in the most intricate manner, but from which fibres are again given off to pursue their independent course, *e. g.*, brachial, cervical, lumbar, sacral, cardiac plexuses, etc.

SPINAL NERVES.

Origin. The spinal nerves are thirty-one in number on each side of the spinal cord, and arise by two roots, an *anterior* and *posterior*, from the anterior and posterior aspects of the cord respectively: the *posterior roots* present near their emergence from the cord a small ganglionic enlargement; outside of the spinal canal the two roots unite to form a main trunk, which is ultimately distributed to the skin, muscles and viscera.

The **Function of the Anterior Roots** is to transmit *motor* impulses from the **centres outward** to the periphery. *Irritation* of these roots, from whatever cause, excites convulsive movements in the muscles to which they are distributed; *disease* or *division* of these roots induces a condition of paresis or paralysis.

The **Function of the Posterior Roots** is to transmit the impressions made upon the periphery to the centres in the spinal cord, where they excite motor impulses; or to the brain, in which they are translated into conscious sensations. *Irritation* of these roots gives rise to painful sensations; *division* of the roots abolishes all sensation in the parts to which they are distributed.

The *ganglion* on the posterior root influences the nutrition of the sensory nerve; for if the nerve be separated from the ganglion, it undergoes degeneration in the course of a few days, in the direction in which it carries impressions, *i. e.*, from the periphery to the centres; if the nerve be *divided between* the ganglion and the cord, the central end only undergoes degeneration. The nutrition of the *anterior root* is governed by nerve cells in the gray matter of the cord; for if these cells undergo *atrophy*, or if the nerve be *divided*, it undergoes degeneration outward.

Nerve Terminations. (1) *Central.* Both motor and sensory nerve fibres, as they enter the spinal cord and brain, lose their external investments, and retaining only the axis cylinder, ultimately become connected with the processes of the gray cells.

(2) *Peripheral.* As the nerves approach the tissues to which they are to be distributed, they inosculate freely, forming a *plexus* from which the ultimate fibres proceed to individual tissues.

Motor Nerves. In the *voluntary* or *striped* muscles the motor nerves are connected with the contractile substance by means of the "*motorial end plates;*" when the nerve enters the muscular fibre the tubular membrane blends with the sarcolemma, the medullary layer disappears, and the axis cylinder spreads out into the form of a little plate, granular in character, and containing oval nuclei.

In the *unstriped* or *involuntary* muscles, the terminal nerve fibres form a plexus on the muscular fibre cells, and become connected with the granular contents of the nuclei.

In the *glands* nerve fibres have been traced to the glandular cells, where they form a branching plexus from which fibres pass into their interior and become connected with their substance, and thus influence secretion.

Sensitive Nerves terminate in the skin and mucous membranes, in three distinct modes, *e. g.*, as tactile corpuscles, Pacinian corpuscles, and as end bulbs.

The *tactile corpuscles* are found in the papillæ of the true skin, especially on the palmar surface of the hands and fingers, feet and toes; they are oblong bodies, measuring about $\frac{1}{300}$ of an inch in length, consisting of a central bulb of homogeneous connective tissue surrounded by elastic fibres and elongated nuclei. The nerve fibre approaches the base of the corpuscle, makes two or three spiral turns around it, and terminates in loops. They are connected with the sense of touch.

The *Pacinian corpuscles* are found chiefly in the subcutaneous cellular tissue, on the nerves of the hands and feet, the intercostal nerves, the cutaneous nerves, and in many other situations. They are oval in shape, measure about the $\frac{1}{16}$ of an inch in length on the average, and consist of concentric layers of connective tissue; the nerve fibre penetrates the corpuscle and terminates in a rounded knob in the central bulb. Their *function* is unknown.

The *end bulbs* of Krause are formed of a capsule of connective tissue in which the nerve fibre terminates in a coiled mass or bulbous extremity; they exist in the conjunctiva, tongue, glans penis, clitoris, etc.

Many *sensitive* nerves terminate in the papillæ at the base of the hair follicle; but in the skin, mucous membranes, and organs of special sense their mode of termination is not well understood.

PROPERTIES AND FUNCTIONS OF NERVES.

Classification. Nerves may be divided into two groups, viz. :—

(1) *Afferent* or *centripetal,* as when they convey to the nerve centres the impressions which are made upon their peripheral extremities or parts of their course. They may be *sensitive,* when they transmit impressions which give rise to sensations; *reflective* or *excitant,* when the impression carried to the nerve centre is reflected outward by an efferent nerve and produces motion or some other effect in the part to which the nerve is distributed.

(2) *Efferent* or *centrifugal,* as when the impulses generated in the centres are transmitted outward to the muscles and various organs. They may be *motor,* as when they convey impulses to the voluntary and involuntary muscles; *vasomotor,* when they regulate the calibre of the small blood vessels, increasing or diminishing the amount of blood to a part; *secretory,* when they influence secretion; *trophic,* when they influence nutrition; *inhibitory,* when they conduct impulses which produce a restraining or inhibiting action.

Stimuli of Nerves. Nerves respond to stimulation according to their habitual function; thus, stimulation of a sensory nerve, if sufficiently strong, results in the sensation of pain; of the optic nerve, in the sensation of light; of a motor nerve, in contraction of the muscle to which it is distributed; of a secretory nerve, in the activity of the related gland, etc. It is, therefore, evident that peculiarity of nervous function depends neither upon any special construction or activity of the nerve itself, nor upon the nature of the stimulus, but entirely upon the peculiarities of its central and peripheral end organs. Nerves react to the following kinds of stimuli :—

1. Mechanical : as from a blow, pressure, tension, puncture, etc.
2. Thermal : heating a nerve at first increases and then decreases its excitability.

3. Chemical : Sensory nerves respond somewhat less promptly than motor nerves to this form of irritatioñ.

4. Electrical : Either the constant or interrupted current.

5. The normal physiological stimulus :—

(a) Centrifugal or efferent if proceeding from the centre toward the periphery.

(b) Centripetal or afferent if in the reverse direction.

(c) Reflex, a combination of the two preceding.

The *Axis Cylinder* is the essential conducting agent, the white substance of Schwann **and** tubular membrane being probably accessory structures, protecting **the axis** from injury, and preventing the diffusion of nerve force to adjoining nerves.

The properties of *sensation* and *motion* reside in different nerve fibres. *Motor* nerves can be destroyed or paralyzed by the introduction of woorara under the skin, without affecting sensation ; the *sensibility* of nerves can be abolished by the employment of anæsthetics without destroying motion.

Irritability. Nerves conduct peripheral impressions to the centres, and motor impulses to the periphery, in virtue of their possessing an ultimate and inherent property, denominated *neurility, nervous irritability*, or *excitability*, which is manifested as long as the physical and chemical integrity of the nerve is maintained.

Nerve Degeneration. When nerves are separated from their trophic or **nutritive** centres, they degenerate progressively in the direction in which they conduct impressions. In *motor* nerves, from the centre to the periphery ; in sensory nerves, from the periphery to the centres.

Nerve force is not identical with electricity. Nerves do not possess the power of generating force, or of originating impulses within themselves, but propagate only the nervous impulses which are called forth by chemical, physical and mechanical stimuli from without, and by volitional acts, normal and pathological conditions from within.

Velocity of Nerve Force. This is slightly modified by temperature, cold lessening the rapidity, heat increasing it ; it is also modified by electrical conditions, by the action of drugs and by the strength of the stimulus. The average velocity for human sensory nerves has been estimated at about 190 feet per second, and for motor nerves 100 to 120 feet per second.

Phenomena of Muscles and Nerves. The muscles are the motor **organs** of the body and constitute a large per cent. of the body weight. Muscles are of two kinds, striated and non-striated or involuntary. The striated muscles consist of bundles of fibres, the *fasciculi*, held together by connective tissue. Each muscle fibre is about ½ to 1½ inches long, and possesses a delicate homogeneous membrane, the *sarcolemma*, in the interior of which is contained the contractile substance, which presents a striated appearance. During life this substance is in a fluid condition, but after death undergoes stiffening.

The non-striated **muscles form** membranes which surround cavities, *e.g.*, stomach, arteries, bladder, etc. They are composed of elongated cells without striations, and contain in their interior one or more nuclei.

Muscular tissue is composed of water, an organic contractile substance, *myosin*, non-nitrogenized substances, such as glycogen, inosite, fat, and inorganic **salts**. When at rest the muscle is alkaline in reaction, but during **and** after contraction it becomes acid.

Muscles possess the properties of (1) *Contractility*, which is the capability of shortening themselves in the direction of their long axis, and at the same time becoming thicker and more rigid. (2) *Extensibility*, by means of which they are lengthened in proportion to weights attached. (3) *Elasticity*, in virtue of which they return to their original shape when the force applied is removed.

The contractility of muscles is called forth mainly by nervous impulses, descending motor nerves, which originate in the central nervous system; but it can also be excited by the electric current, the application of strong acids, heat, or by mechanical means.

Phenomena of a Muscular Contraction. When **a single induction** shock is propagated through a nerve, the muscle to which it is distributed undergoes a quick pulsation, and speedily **returns to its former condition.** As is shown by the muscle curve, the contraction, **which is at first slow,** **increases in rapidity** to its maximum, gradually **relaxes and** is again at rest, the entire pulsation not occupying more than the $\frac{1}{10}$ of a second.

The **muscular** contraction does not instantly follow the induction shock, even when **the** electrodes are placed directly upon the muscular fibres themselves; an appreciable period intervenes before the contraction, during which certain chemical changes are taking place preparatory to the manifestation of force. This is the " latent period," which has an average duration of the $\frac{1}{100}$ of a second, but varies with the temperature, the strength of the stimulus, the animal, etc. The muscular movements of the body, however, are occasioned by contractions of a much longer duration, depending upon the number (the average, 20) of nervous impulses passing to the muscles in a second.

During the muscular contraction the following phenomena are observed, viz.: a change in form, a rise in temperature, a consumption of oxygen and an evolution of carbonic acid; the production of a distinct musical sound,

G

a change from an alkaline to an acid reaction, from the development of
sarcolactic acid; a disappearance of the **natural** muscle currents, which
undergo a *negative variation* in the "latent period," **just** after the nervous
impulse reaches the termination of the nerve, and before the appearance
of the muscular contraction wave.

Electrical Properties of Nerves. When a galvanic current is made
to flow along a motor nerve from the centre to the periphery, from the
positive to the negative pole, it is known as the *direct, descending* or *centrifugal* current. When it is made to flow in the reverse direction it is known
as the *inverse, ascending* or *centripetal* current.

The passage of a direct current enfeebles the excitability of a nerve; the
passage of the inverse current increases it. The excitability of a nerve
may be exhausted by the repeated applications of electricity; when thus
exhausted it may be restored by repose, or by the passage of the inverse
current if the nerve has been exhausted by the direct current or vice versa.

During the actual passage of a feeble constant current in either direction
neither pain nor muscular contraction is ordinarily manifested; if the current be very intense the nerve may be disorganized and its excitability
destroyed.

Electrotonus. The passage of **a direct** galvanic current through a portion of a nerve excites in the parts beyond the electrodes a condition of
electric tension or *electrotonus*, during which the *excitability* of the nerve
is *decreased* near the anode or positive pole, and *increased* near the kathode
or negative pole; the increase of excitability in the *kathelectrotonic area*,
that nearest the muscle, being manifested by a more marked contraction of
the muscle than the normal, when the nerve is irritated in this region. The
passage of an *inverse* galvanic current excites the same condition of
electrotonus; and the *diminution* of excitability near the anode, the *anelectrotonic area*, that now nearest the muscle, being manifested by a less
marked contraction than the normal when the nerve is stimulated in this
region. Between the electrodes is a neutral point where the kathelectrotonic area emerges into the anelectrotonic area. If the current be a strong
one, the neutral point approaches the kathode; if weak, it approaches the
anode.

When a nervous impulse passes along a nerve, the only appreciable
effect is a change in its electrical condition, there being no change in its
temperature, chemical composition or physical condition. The natural
nerve currents, which are always present in a living nerve as a result of its
nutritive activity, in great part disappear during the passage of an impulse,
undergoing a *negative variation*.

Law of Contraction. If a *feeble* galvanic current be applied to a recent and excitable nerve, contraction is produced in the muscles only upon the *making* of the circuit with both the direct and inverse currents.

If the current be *moderate* in intensity, the contraction is produced in the muscle both upon the *making* and *breaking* of the circuit, with both the direct and inverse currents.

If the current be *intense*, contraction is produced only when the circuit is *made* with the direct current, and only when it is *broken* with the inverse current.

The Reaction of Degeneration. Two different applications of electricity are used in electro-physiology and electro-therapeutics—the constant or galvanic, and the interrupted or faradic currents. Injured and paralyzed muscles and nerves react differently to these two kinds of stimuli, and the facts are of the greatest importance in the diagnosis and therapeutics of the precedent lesions. The principal difference of behavior relates to the *reaction of degeneration*—a condition produced by paralysis of any kind. It is characterized by a diminished or abolished excitability of the muscles to the faradic current, while there is at the same time an increased excitability to the galvanic current. The synchronous diminished excitability of the nerves is the same for either current. The term partial reaction of degeneration is used when there is a normal reaction of the nerves, but the muscles show the degenerative reaction. This condition is a characteristic of progressive muscular atrophy.

CRANIAL NERVES.

The Cranial Nerves come off from the base of the brain, pass through the foramina in the walls of the cranium, and are distributed to the skin, muscles and organs of sense in the face and head.

According to the classification of Sœmmering, there are 12 pairs of nerves, enumerating them from before backward, as follows, viz.:—

1st Pair, or Olfactory.	7th Pair, or Facial, Portio dura.
2d Pair, or Optic.	8th Pair, or Auditory, Portio mollis.
3d Pair, or Motor oculi communis.	9th Pair, or Glosso-pharyngeal.
4th Pair, or Patheticus, Trochlearis.	10th Pair, or Pneumogastric.
5th Pair, or Trifacial, Trigeminus.	11th Pair, or Spinal accessory.
6th Pair, or Abducens.	12th Pair, or Hypoglossal.

The Cranial Nerves may also be classified physiologically, according to their function, into three groups: 1. Nerves of special sense. 2. Nerves of motion. 3. Nerves of general sensibility.

1st Pair. Olfactory.

Apparent Origin. From the inferior and internal portion of the anterior lobes of the cerebrum by three roots, viz.: an *external white root*, which passes across the fissure of Sylvius to the middle lobe of the cerebrum; an *internal white root*, from the most posterior part of the **anterior lobe**; a *gray root*, from the gray matter in the posterior and inner portion of the inferior surface of the anterior lobe.

Deep Origin. Not satisfactorily determined.

Distribution. The olfactory nerve, formed by the union of the three roots, passes forward along the under surface of the anterior lobe to the ethmoid bone, where it expands into the olfactory bulb. This bulb contains ganglionic cells, is grayish in color and soft in consistence; it gives off from its under surface from fifteen to twenty nerve filaments, the *true olfactory nerves*, which pass through the cribriform plate of the ethmoid bone, and are distributed to the schneiderian mucous membrane. This membrane extends from the cribriform plate of the ethmoid bone downward, about one inch.

Properties. The olfactory nerves give rise to neither motor nor sensory phenomena when stimulated. They carry simply the special impressions of odorous substances. Destruction or injury of the olfactory bulbs is attended by a loss of the sense of smell.

Function. Governs the sense of smell. Conducts the impressions which give rise to odorous sensations.

2d Pair. Optic.

Apparent Origin. From the anterior portion of the optic commissure.

Deep Origin. The origins and connections of the optic tract are very complex. The immediate origins are bands of fibres from the thalamus opticus and anterior corpora quadrigemina. The corpora geniculata are interposed ganglia. The ultimate roots are traced—

1. By a broad band of fibres—"the optic radiation of Gratiolet"—to the psycho-optic centres in the occipital lobes.
2. To the gyrus hippocampi and sphenoidal lobes.
3. Through the corpus callosum to the motor areas of the opposite cerebral hemispheres.
4. To the frontal region by "Meynert's Commissure."
5. To the spinal cord.
6. To the corpora geniculata, pulvinar, and anterior corpora geniculata by ganglionic roots.

Distribution. The two roots unite to form a flattened band, the optic tract, which winds around the crus cerebri to decussate with the nerve of the opposite side, forming the *optic chiasm.* The decussation of fibres is not complete; some of the fibres of the left optic tract going to the outer half of the eye of the same side, and to the inner half of the eye of the opposite side; the same holds true for the right optic tract.

The *optic nerves proper* arise from the commissure, pass forward through the optic foramina, and are finally distributed in the *retina.*

Properties. They are insensible to ordinary impressions, and convey only the *special impressions of light.*

Division of one of the nerves is attended by complete blindness in the eye of the corresponding side; *division* of the *optic tract* produces *loss* of *sight* in the *outer half* of the eye of the same side, and in the *inner half* of the eye of the opposite side. Lesion of the anterior part of the *optic chiasm* causes blindness in the inner half of the two eyes.

Functions. Governs the sense of sight. Receives and conveys to the brain the luminous impressions which give rise to the sensation of sight.

The reflex movements of the iris are called forth by the optic nerve. When an excess of light falls upon the retina the impression is carried back to the tubercula quadrigemina, where it is transformed into a motor impulse, which then passes outward through the motor oculi nerve to the contractile fibres of the iris and diminishes the size of the pupil. The absence of light is followed by a dilatation of the pupil.

3d Pair. Motor Oculi Communis.

Apparent Origin. From the inner surface of the crura cerebri.

Deep Origin. By filaments coming from the lenticular nucleus, corpora quadrigemina, optic thalamus; these filaments converge to form a main trunk, which winds around the crus cerebri, in front of the pons Varolii.

Distribution. The nerve then passes forward, and enters the orbit through the sphenoidal fissure, where it divides into a *superior branch* distributed to the *superior rectus* and *levator palpebræ* muscles; an *inferior* branch sending branches to the *internal* and *inferior recti,* and the *inferior oblique* muscles; filaments also pass into the *ciliary* or *ophthalmic* ganglion; from this ganglion the *ciliary nerves* arise which enter the eyeball, and are distributed to the *circular fibres of the iris* and the *ciliary muscle.* The 3d nerve also receives filaments from the cavernous plexus of the sympathetic and from the fifth nerve.

Properties. *Irritation* of the root of the nerve produces contraction of the pupil, internal strabismus, muscular movements of eye, but no pain. *Division* of the nerve is followed by *ptosis* (falling of the upper eyelid), *external strabismus*, due to the unopposed action of the external rectus muscle; paralysis of the accommodation of the eye; *dilatation* of the pupil from paralysis of the circular fibres of the iris and ciliary **muscle**; and *inability* to *rotate* the eye, *slight protrusion* and *double vision*. **The** images are crossed; that of the paralyzed eye is a little above that **of the** sound, and its upper end inclined toward it.

Function. Governs movements of the eyeball by animating all the muscles except the external rectus and superior oblique, the movements of the iris, elevates the upper lid, influences the accommodation of the eye for distances. Can be called into action by (1) voluntary stimuli, (2) by reflex action through irritation of the optic nerve.

4th Pair. Patheticus.

Apparent Origin. From the superior peduncles of the cerebellum.

Deep Origin. By fibres terminating in the corpora quadrigemina, lenticular nucleus, valve of Vieussens, and in the substance of the cerebellar peduncles; some filaments pass over the median line and decussate with fibres of the opposite side.

Distribution. The nerve enters the orbital cavity through the sphenoidal fissure, and is distributed to the *superior oblique* muscle; in its course receives filaments from the ophthalmic branch of the 5th pair and the **sympathetic.**

Properties. When the nerve is *irritated* muscular movements **are produced** in the superior oblique muscle, and the pupil of the eye is turned *downward* and *outward*. Division or paralysis lessens the movements and rotation of the globe downward and outward. The diplopia consequent upon this paralysis is homonymous, one image appearing above the other. The image of the paralyzed eye is below, its upper end inclined toward that of the sound eye.

Function. Governs the movements of the eyeball produced by the action of the superior oblique muscles.

6th Pair.* Abducens. Motor Oculi Externus.

Apparent Origin. From the groove between the anterior pyramidal body and the pons Varolii, where it arises by two roots.

---

* The 6th nerve is considered in connection with the 3d and 4th nerves, since they **together constitute** the motor apparatus by which the ocular muscles are excited to **action.**

Deep Origin. From the gray matter of the medulla oblongata.

Distribution. The nerve then passes into the orbit through the sphenoidal fissure, and is distributed to the *external rectus* muscle. Receives filaments from the cervical portion of the sympathetic, through the carotid plexus and spheno-palatine ganglion.

Properties. When *irritated*, the *external rectus muscle* is thrown into convulsive movements, and the eyeball is turned outward. When *divided* or *paralyzed*, this muscle is paralyzed; motion of the eyeball outward past the median line is impossible, and the homonymous diplopia increases as the object is moved outward past this line. The images are upon the same plane and parallel. Internal strabismus results because of the unopposed action of the internal rectus.

Function. To turn the eyeball outward.

5th Pair. Trifacial. Trigeminal.

Apparent Origin. By two roots from the side of the pons Varolii.

Deep Origin. The deep origin of the two roots is the upper part of the floor and anterior wall of the 4th ventricle, by three bundles of filaments, one of which anastomoses with the auditory nerve; another passes to the lateral tract of the medulla; while a third, grayish in color, goes to the restiform bodies, and may be traced to the point of the calamus scriptorius.

Filaments of origin have been traced to the "trigeminal sensory nucleus," located on a level with the point of exit of the nerve, and to the posterior gray horns of the cord, as low down as the middle of the neck.

Distribution. The *large root* of the nerve passes obliquely upward and forward to the ganglion of Gasser, which receives filaments of communication from the carotid plexus of the sympathetic. It then divides into three branches.

1. *Ophthalmic branch*, which receives communicating filaments from the sympathetic, and sends sensitive fibres to all the motor nerves of the eyeball. It is *distributed* to the ciliary ganglion, lachrymal gland, sac and caruncle, conjunctiva, integument of the upper eyelid, forehead, side of head and nose, anterior portion of the scalp, ciliary muscle and iris.

2. *Superior maxillary branch*, sends branches to the spheno-palatine ganglion, integument of the temple and lower eyelid, side of forehead, nose, cheek and upper lip, teeth of the upper jaw, and alveolar processes.

3. *Inferior maxillary branch*, which, after receiving in its course filaments from the *small root* and from the *facial*, is distributed to the sub-

maxillary ganglion, the parotid and sub-lingual glands, external auditory meatus, mucous membrane of the mouth, anterior two-thirds of the tongue (lingual branch), gums, arches of the palate, **teeth of** the lower jaw, and integument of the lower part of the face, and *to the muscles of mastication*.

The *small root* passes forward beneath the ganglion of Gasser, through the foramen ovale, and joins the *inferior maxillary* division of the large root, which then divides into an anterior and posterior branch, the former of which is distributed to the muscles of mastication, viz.: temporal, masseter, internal and external pterygoid muscles.

Properties. It is the most acutely sensitive **nerve** in the body, and endows all the parts to which it is distributed with general sensibility.

Irritation of the *large root*, or any of its branches, will give rise to **marked evidence of** pain; the various forms of neuralgia **of the** head and **face** being occasioned by compression, disease, or exposure **of** some of its terminal branches.

Division of the large root within the cranium is followed at once by a complete abolition of all sensibility in the head and face, but is not attended by any loss of motion. The integument, mucous membranes and the eye may be lacerated, cut or bruised, without the animal exhibiting any evidence of pain. At the same time the lachrymal secretion is diminished, the pupil becomes contracted, the eyeball is protruded, and the sensibility of the tongue is abolished.

The reflex movements of deglutition are also somewhat impaired; the impression of the food being unable to reach and excite the nerve centre in the medulla oblongata.

Galvanization of the *small root* produces movements of the muscles of mastication; *section* of the root causes paralysis of these muscles, and the jaw is drawn to the opposite side, by the action of the opposing muscles.

Influence upon the Special Senses. After division of the large root within the cranium, a disturbance in the nutrition of the special senses sooner or later manifests itself.

Sight. In the course of twenty-four hours the *eye* becomes very vascular and inflamed, the cornea becomes opaque and ulcerates, the humors are discharged, and the eye is totally destroyed.

Smell. The nasal mucous membrane swells up, becomes fungous, and is liable to bleed on the slightest irritation. The mucus is increased in amount, so as to obstruct the nasal passages; the sense of smell is finally abolished.

Hearing. At times the hearing is impaired, from disorders of nutrition in the middle ear and external auditory meatus.

Alteration in the nutrition of the special senses is not marked if the section is made posterior to the ganglion of Gasser, and to the anastomosing filaments of the sympathetic which join the nerve at this point; but if the ganglion be divided, these effects are very noticeable, due to the section of the sympathetic filaments.

Function. Gives sensibility to all parts of the head and face to which it is distributed; through the small root endows the masticatory muscles with motion; through fibres from the sympathetic governs the nutrition of the special senses.

7th Pair. Portio Dura. Facial Nerve.

Apparent Origin. From the groove between the olivary and restiform bodies at the lateral portion of the medulla oblongata, and below the margin of the pons Varolii.

Deep Origin. From a nucleus of large cells in the floor of the 4th ventricle, below the nucleus of origin of the 6th pair, with which it is connected. Some filaments are traceable to the lenticular nucleus of the opposite side. Some of the fibres cross the median line and decussate. It is intimately associated with the nerve of Wrisberg at its origin.

Distribution. From its origin the facial nerve passes into the internal auditory meatus, and then, in company with the nerve of Wrisberg, enters the aqueduct of Fallopius. The filaments of the nerve of Wrisberg are supplied with a ganglion, of a reddish color, having nerve cells. These filaments unite with those of the root of the facial, to form a common trunk, which emerges at the stylo-mastoid foramen.

In the aqueduct the facial gives off the following branches, viz.:—

1. *Large petrosal nerve*, which passes forward to the *spheno-palatine*, or Meckel's ganglion, and through this to the levator palati and azygos uvulæ muscles, which receive motor influence from this source.

2. *Small petrosal nerve*, passing to the *otic* ganglion and thence to the *tensor-tympani* muscle, endowing it with motion.

3. *Tympanic branch*, giving motion to the stapedius muscle.

4. *Chorda tympani* nerve, which after entering the posterior part of the tympanic cavity, passes forward between the malleus and incus bones, through the Glasserian fissure, and joins the lingual branch of the 5th nerve. It is then distributed to the mucous membrane of the anterior two-thirds of the tongue and the sub-maxillary glands.

After emerging from the stylo-mastoid foramen, the facial nerve sends branches to the muscles of the ear, the occipito-frontalis, the digastric, the palato-glossi, and palato-pharyngei; after which it passes through the parotid gland and divides into the *temporo-facial* and *cervico-facial* branches, which are distributed to the superficial muscles of the face, viz.: occipito-frontalis, corrugator supercilii, orbicularis palpebrarum, levator labii superioris et alæque nasi, buccinator, levator anguli oris, orbicularis oris, **zygomatici**, depressor anguli oris, platysma myoides, etc.

Properties. Undoubtedly a motor nerve at its origin, but in its course receives sensitive filaments from the 5th pair and the pneumogastric.

Irritation of the nerve, after its emergence from the stylo-mastoid foramen, produces convulsive movements in all the superficial muscles of the face. *Division* of the nerve at this point causes paralysis of these muscles on the side of the section, constituting *facial paralysis;* the phenomena of which are, a relaxed and immobile condition of the same side of the face; the eyelids remain open, from paralysis of the orbicularis palpebrarum; the act of winking is abolished; the angle of the mouth droops, and saliva constantly drains away; the face is drawn over to the sound side; the face becomes distorted upon talking or laughing; mastication is interfered with, the food accumulating between the gums and cheek, from paralysis of the buccinator muscle; fluids escape from the mouth in drinking; **articulation** is impaired, the labial sounds being imperfectly pronounced.

Properties of the branches given off in the aqueduct of Fallopius. The *Large petrosal*, when irritated, throws the levator palati and **azygos uvulæ** muscles into contraction. *Paralysis* of this nerve, from deep-seated lesions, produces a *deviation* of the uvula to the sound side, a *drooping* of the palate, and an inability to elevate it.

The *Small petrosal* influences hearing by animating the tensor tympani muscle; when paralyzed, there occurs partial deafness and an increased sensibility to sonorous impressions.

The *Tympanic branch* animates the stapedius muscle, and influences audition.

The *Chorda tympani* influences the circulation and the secretion of saliva, in the sub-maxillary glands, and governs the sense of taste in the anterior two-thirds of the tongue. *Galvanization* of the chorda tympani dilates the blood vessels, increases the quantity and rapidity of the stream of blood, and increases the secretion of saliva. *Division* of the nerve is followed by contraction of the vessels, an arrestation of the secretion, and a diminution of the sense of taste, on the same side.

Function. The facial is the nerve of expression, and coördinates the muscles employed to delineate the various emotions, influences the sense of taste, deglutition, movements of the uvula and soft palate, the tension of the membrana tympani, and the secretions of the sub-maxillary and parotid glands. Indirectly influences smell, hearing and vision.

8th Pair. Portio Mollis. Auditory Nerve.

Apparent Origin. From the upper and lateral portion of the medulla oblongata, just below the margin of the pons Varolii.

Deep Origin. By two roots from the floor of the 4th ventricle, each root consisting of a number of gray filaments, some of which decussate in the median line; the external root has a gangliform enlargement containing fusiform nerve cells.

Distribution. The two roots wind around the restiform bodies and enter the internal auditory meatus, and divide into an anterior branch distributed to the cochlea, and a posterior branch distributed to the vestibule and semicircular canals.

Properties. They are soft in consistence, grayish in color, consisting of axis cylinders with a medullary sheath only; they are not sensible to ordinary impressions, but convey the *impression of sound.*

Function. Governs the sense of hearing. Receives and conducts to the brain the impression of sound, which gives rise to the sensations of hearing.

9th Pair. Glosso-pharyngeal.

Apparent Origin. Partly from the medulla oblongata and the inferior peduncles of the cerebellum.

Deep Origin. From the lower portion of the gray substance in the floor of the 4th ventricle.

This nerve has two ganglia; the *jugular ganglion* includes only a portion of the root filaments; the ganglion of Andersch includes all the fibres of the trunk.

Distribution. The trunk of the nerve passes downward and forward, receiving near the ganglion of Andersch fibres from the facial and pneumogastric nerves. It divides into two large branches, one of which is distributed to the base of the tongue, the other to the pharynx. In its course it sends filaments to the otic ganglion; a tympanic branch which gives sensibility to the mucous membrane of the fenestra rotunda, fenestra ovalis, and Eustachian tube; lingual branches to the base of the tongue; palatal branches to the soft palate, uvula and tonsils; pharyngeal branches to the mucous membrane of the pharynx.

Properties. *Irritation* of the roots at their origin calls forth evidences of pain; it is, therefore, a sensory nerve, but its sensibility is not so acute as that of the trifacial. *Irritation* of the trunk after its exit from the cranium produces contraction of the muscles of the palate and pharynx, due to the presence of anastomosing motor fibres.

Division of the nerve abolishes sensibility in the structures to which it is distributed, and impairs the sense of taste in the posterior third of the tongue (see Sense of Taste).

Function. Governs sensibility of pharynx, presides partly over the sense of taste, and controls reflex movements of deglutition and vomiting.

10th Pair. Pneumogastric. Par Vagum

Apparent Origin. From the lateral side of the medulla oblongata, just behind the olivary body.

Deep Origin. In the gray nuclei in the lower half of the floor of the 4th ventricle, and in the substance of the restiform body. Some filaments are traced along the restiform tract, toward the cerebellum, and others to the median line of the floor of the 4th ventricle, where many of them decussate.

This nerve has two ganglia; one in the jugular foramen, called the ganglion of the root, and another outside of the cranial cavity on the trunk, the ganglion of the trunk.

Distribution. The filaments from the root unite to form a single trunk, which leaves the cavity of the cranium, through the jugular foramen, in company with the spinal accessory and glosso-pharyngeal. It soon receives an *anastomotic branch* from the *spinal accessory*, and afterward branches from the facial, the hypoglossal and the anterior branches of the two upper cervical nerves.

As the nerve passes down the neck it sends off the following main branches:—

1. *Pharyngeal nerves*, which assist in forming the pharyngeal plexus, which is distributed to the mucous membrane and muscles of the pharynx.

2. *Superior laryngeal nerve*, which enters the larynx through the thyro-hyoid membrane, and is distributed to the mucous membrane lining the interior of the larynx, and to the crico-thyroid muscle and the inferior constrictor of the pharynx. The "*depressor nerve*," found in the rabbit, is formed by the union of two branches, one from the superior laryngeal, the other from the main trunk; it passes downward to be distributed to the heart.

3. *Inferior laryngeal*, which sends its ultimate branches to all the intrinsic muscles of the larynx except the crico-thyroid, and to the inferior constrictor of the pharnyx.

4. *Cardiac* branches given off from the nerve throughout its course, which unite with the sympathetic fibres to form the cardiac plexus, to be distributed to the heart.

5. *Pulmonary branches*, which form a plexus of nerves and are distributed to the bronchi and their ultimate terminations, the lobules and air cells.

From the right pneumogastric nerve branches are distributed to the mucous membrane and muscular coats of the stomach and intestines, to the liver, spleen, kidneys, and supra-renal capsules.

Properties. At its origin the pneumogastric nerve is sensory, as shown by direct irritation or galvanization, though its sensibility is not very marked. In its course exhibits motor properties, from anastomosis with motor nerves.

The *Pharyngeal branches* assist in giving sensibility to the mucous membrane of the pharynx, and influence reflex phenomena of deglutition through motor fibres which they contain, derived from the spinal accessory.

The *Superior laryngeal* nerve endows the upper portion of the larynx with sensibility; protects it from the entrance of foreign bodies; by conducting impressions to the medulla, excites the reflex movements of deglutition and respiration; through the motor filaments it contains produces contraction of the crico-thyroid muscle.

Division of the " *Depressor nerve*," and *galvanization* of the central end, retards and even arrests the pulsations of the heart, and by depressing the vasomotor centre diminishes the pressure of blood in the large vessels, by causing dilatation of the intestinal vessels through the splanchnic nerves.

The *Inferior laryngeal* contains, for the most part, motor fibres from the spinal accessory. When *irritated*, produces movement in the laryngeal muscles. When *divided*, is followed by paralysis of these muscles, except the crico-thyroid, impairment of phonation, and an embarrassment of the respiratory movements of the larynx, and finally death, from suffocation.

The *Cardiac branches*, through filaments derived from the spinal accessory, exert a direct inhibitory action upon the heart. *Division* of the pneumogastrics in the neck increases the frequency of the heart's action. *Galvanization* of the peripheral ends diminishes the heart's pulsation, and, if sufficiently powerful, paralyzes it in diastole.

The *Pulmonary branches* give sensibility to the bronchial mucous membrane, and govern the movements of respiration. *Division* of both pneumogastrics in the neck diminishes the frequency of the respiratory movements, falling as low as four to six per minute; death usually occurs in from five to eight days. *Feeble galvanization* of the central ends of the divided nerves accelerates respiration; *powerful galvanization* retards, and may even arrest the respiratory movements.

The *Gastric branches* give sensibility to the mucous coat, and **through** sympathetic filaments, which join the pneumogastrics high up in the neck, give motion to the muscular coat of the stomach. They influence the secretion of gastric juice, aid the process of digestion and absorption from **the stomach**.

The *Hepatic branches,* probably through anastomosing sympathetic filaments, influence the secretion of bile, and the glycogenic function of the liver; *division* of the pneumogastrics in the neck produces congestion of **the** liver, diminishes the density of the bile, and arrests the glycogenic function; *galvanization* of the central ends exaggerates the glycogenic function, and makes the animal diabetic.

The *Intestinal branches* give sensibility and motion to the small intestines, and when divided, purgatives generally fail to produce purgation.

Function. A great sensitive nerve, which, through anastomotic filaments from motor sources, influences deglutition, the action of the heart, the circulatory and respiratory systems, voice, the secretions of the stomach, intestines, and various glandular organs.

11th Pair. Spinal Accessory.

Apparent Origin. By two sets of filaments : —

1. A bulbar **or** medullary set, four or five in number, from the lateral or motor tract of the lower half of the medulla oblongata, below the origin of the pneumogastric.

2. A spinal set, from six to eight in number, from the lateral portion of the spinal cord, between the anterior and posterior roots of the upper four or five cervical nerves.

Deep Origin. The *medullary portion* arises in a nucleus in the lower half of the floor of the 4th ventricle, common to the pneumogastric and glosso-pharyngeal nerves. The *spinal portion* has its origin in an elongated nucleus lying along the external surface of the anterior cornua of the spinal cord, extending down to the 5th cervical vertebra.

Distribution. From this origin the fibres unite to form a main trunk, which enters the cranial cavity through the foramen magnum, where it is at times joined by fibres from the posterior roots of the two upper cervical nerves, and sends filaments to the ganglion of the root of the pneumogastric. After emerging from the cranial cavity through the jugular foramen, it sends a branch to the pneumogastric, and receives others in return, and also from the 2d, 3d and 4th cervical nerves. It divides into two branches: (1) An *internal* or *anastomotic* branch, made up of filaments coming principally from the medulla oblongata, and is distributed to the muscles of the pharynx through the pharyngeal nerves coming from the pneumogastric; to all the muscles of the larynx, except the crico-thyroid through the *inferior laryngeal* nerve; to the heart, by filaments which reach it through the pneumogastric nerve. (2) An *external branch*, which is distributed to the sterno-cleido-mastoid and trapezius muscles; these muscles also receiving filaments from the cervical nerves.

Properties. At its origin it is a purely *motor* nerve, but in its course exhibits some sensibility from anastomosing fibres.

Destruction of the *medullary root*, by tearing it from its attachment by means of forceps, impairs the action of the muscles of deglutition, and destroys the power of producing vocal sounds by paralysis of the laryngeal muscles, without, however, interfering with the respiratory movements of the larynx; these being controlled by other motor nerves. The normal rate of movement of the heart is also impaired by destruction of the medullary root.

Irritation of the external branch throws the trapezius and sterno-mastoid muscles into convulsive movements, though *section* of the nerve does not produce complete paralysis, as they are also supplied with motor influence from the cervical nerves. The sterno-mastoid and trapezius muscles perform movements antagonistic to those of respiration, fixing the head, neck and upper part of the thorax, and delaying the expiratory movement during the acts of pushing, pulling, straining, etc., and in the production of a prolonged vocal sound, as in singing. When the *external* branch alone is divided, in animals, they experience shortness of breath during exercise, from a want of coördination of the muscles of the limbs and respiration; and while they can make a vocal sound, it cannot be prolonged.

Function. Governs phonation by its influence upon the vocal movements of the glottis; influences the movements of deglutition, inhibits the action of the heart and controls certain respiratory movements associated with sustained or prolonged muscular efforts and phonation.

12th Pair. Hypoglossal or Sublingual.

Apparent Origin. By two groups of filaments from the medulla oblongata, in the grooves between the olivary body and the anterior pyramid.

Deep Origin. From the hypoglossal nucleus situated deeply in the substance of the medulla, on a level with the lowest portion of the floor of the 4th ventricle; some decussating filaments have been traced to a higher encephalic centre.

Distribution. The trunk formed by a union of the root filaments passes out of the cranial cavity through the anterior condyloid foramen, occasionally receiving a filament from the lateral and posterior portion of the medulla oblongata. After emerging from the cranium, it sends filaments to the sympathetic and pneumogastric; it anastomoses with the lingual branch of the 5th pair, and receives and sends filaments to the upper cervical nerves. The nerve is finally distributed to the sterno-hyoid, sterno-thyroid, omo-hyoid, thyro-hyoid, stylo-glossi, hyo-glossi, genio-hyoid, genio-hyo-glossi, and the intrinsic muscles of the tongue.

Properties. A purely *motor* nerve at its origin, but derives sensibility outside the cranial cavity, from anastomosis with the cervical, pneumogastric and 5th nerves.

Irritation of the nerve gives rise to convulsive movements of the tongue and slight evidences of sensibility.

Division of the nerve abolishes all movements of the tongue, and interferes considerably with the act of deglutition.

When the hypoglossal nerve is involved in hemiplegia, the tip of the tongue is directed to the paralyzed side when the tongue is protruded; due to the unopposed action of the genio-hyo-glossus on the sound side.

Articulation is considerably impaired in paralysis of this nerve; great difficulty being experienced in the pronunciation of the consonantal sounds.

Mastication is performed with difficulty, from inability to retain the food between the teeth until it is completely triturated.

Function. Governs all the movements of the tongue and influences the functions of mastication, deglutition and articulate language.

CEREBRO-SPINAL AXIS.

The Cerebro-Spinal Axis consists of the spinal cord, medulla oblongata, pons Varolii, cerebellum and cerebrum, exclusive of the spinal and cranial nerves. It is contained within the cavities of the cranium and spinal column, and surrounded by three membranes, the dura mater, arachnoid and pia mater, which protect it from injury and supply it with blood vessels.

The Brain and Spinal Cord are composed of both white fibres and collections of gray cells, and are, therefore, to be regarded as conductors of impressions and motor impulses, as well as generators of nerve force.

MEMBRANES.

The Dura Mater, the most external of the three, is a tough membrane, composed of white fibrous tissue, arranged in bundles, which interlace in every direction. In the cranial cavity it lines the inner surface of the bones, and is attached to the edge of the foramen magnum; sends processes inward, forming the falx cerebri, falx cerebelli, and tentorium cerebelli, supporting and protecting parts of the brain. In the spinal canal it loosely invests the cord, and is separated from the walls of the canal by areolar tissue.

The Arachnoid, the middle membrane, is a delicate serous structure which envelopes the brain and cord, forming the *visceral layer*, and is then reflected to the inner surface of the dura mater, forming the *parietal layer*. Between the two layers there is a small quantity of fluid which prevents friction by lubricating the two surfaces.

The Pia Mater, the most internal of the three, composed of areolar tissue and blood vessels, covers the entire surface of the brain and cord, to which it is closely adherent, dipping down between the convolutions and fissures. It is exceedingly vascular, sending small blood vessels some distance into the brain and cord.

The Cerebro-spinal Fluid occupies the *sub-arachnoid space*, and the general ventricular cavities of the brain, which communicate by an opening, the foramen of Magendie, in the pia mater, at the lower portion of the 4th ventricle. This fluid is clear, transparent, alkaline, possesses a salt taste and a low specific gravity; it is composed largely of water, traces of albumen, glucose and mineral salts. It is secreted by the pia mater; the quantity is estimated from two to four fluid ozs.

The *function* of the cerebro-spinal fluid is to protect the brain and cord, by preventing concussion from without; by being easily displaced into the spinal canal, prevents undue pressure and insufficiency of blood to the brain.

H

SPINAL CORD.

The Spinal Cord varies from 16 to 18 inches in length; is half an inch in thickness, weighs 1½ oz., and extends from the atlas to the 2d lumbar vertebra, terminating in the *filum terminale.* It is cylindrical in shape, and presents an enlargement in the lower cervical and lower dorsal regions, corresponding to the origin of the nerves which are distributed to the upper and lower extremities. The cord is divided into two lateral halves by the anterior and posterior fissures. It is composed of both *white* or *fibrous* and *gray* or *vesicular* matter, the former occupying the exterior of the cord, the latter the interior, where it is arranged in the form of two crescents, one in each lateral half, united together by the central mass, the *gray commissure;* the white matter being united in front by the *white commissure.*

Structure of the White Matter. The white matter surrounding each lateral half of the cord is made up of nerve fibres, some of which are continuations of the nerves which enter the cord, while others are derived from different sources. It is subdivided into: (1) An *Anterior* column, comprising that portion between the anterior roots and the anterior fissure, which is again subdivided into two parts: (*a*) an *inner* portion, bordering the anterior median fissure, the *direct pyramidal tract,* or column of Türck, containing motor fibres which do not decussate, and which extends as far down as the middle of the dorsal region; (*b*) an *outer* portion, surrounding the anterior cornua, known as the *anterior root zone,* composed of short longitudinal fibres which serve to connect together different segments of the spinal cord. (2) A *Lateral* column, the portion between the anterior and posterior roots, which is divisible into (*a*) the *crossed pyramidal tract,* occupying the posterior portion of the lateral column, and containing all those fibres of the motor tract which have decussated at the medulla oblongata; it is composed of longitudinally running fibres which are connected with the multipolar nerve cells of the anterior cornua; (*b*) the *direct cerebellar tract,* situated upon the surface of the lateral column, consisting of longitudinal fibres which terminate in the cerebellum; it first appears in the lumbar region, and increases as it passes upward; (*c*) the anterior tract, lying just posterior to the anterior cornua. (3) A *Posterior* column, the portion included between the posterior roots and the posterior fissure, also divisible into two portions, (*a*) an *inner* portion, the *postero-internal column,* or the column of Goll, bordering the posterior median fissure, and (*b*) an *external* portion, the *postero-external column,* the column of Burdach, lying just behind the posterior roots. They are composed of long and short commissural fibres which connect together different segments of the spinal cord.

Structure of the Gray Matter. The gray matter, arranged in the form of two crescents, presents an *anterior* and *posterior horn.* It is made up of a delicate network of fine nerve fibres (axis cylinders), supported by a connective tissue framework of nucleated nerve cells, which in the anterior horns are large and multipolar, and connected with the anterior roots of spinal nerves; in the posterior horns the nerve cells are smaller, and situated along the inner margin,and in the *caput cornu.* Small cells are also found in the *posterior vesicular* columns, and in the intermediary lateral tract.

FIG. 10.

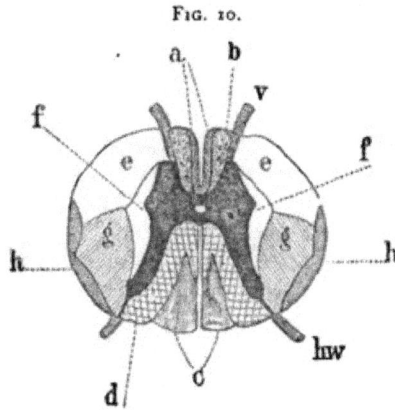

Scheme of the conducting paths in the spinal cord at the 3d dorsal nerve. The black part is the gray matter. v, anterior, h, w, posterior, root; a, direct, and g, crossed, pyramidal tracts; b, anterior column, ground bundle; c, Goll's column; d, postero-external column; e and f, mixed lateral paths; h, direct cerebellar tracts (*Landois*).

COURSE OF THE ANTERIOR AND POSTERIOR ROOTS.

The **Anterior Roots** pass through the anterior columns, horizontally, in straight and distinct bundles, and enter the anterior cornuæ, where they diverge in four directions. (1) Many become connected with the prolongations of the multipolar nerve cells. (2) Others leave the gray matter, pass through the anterior white commissure, and enter the anterior columns of the opposite side. (3) A considerable number enter the *lateral* columns of the same side, through which they pass to the medulla oblongata, where they decussate and finally terminate in the *corpus striatum* of the opposite side. (4) Others traverse the gray matter horizontally, and come into relation with the posterior roots.

The Posterior **Roots** enter the posterior horns of the gray matter (1) *through* the substantia gelatinosa, (2) through the posterior columns; of the *former,* some bend upward and downward, and become connected with the anterior cornuæ; others pass through the posterior commissure to the opposite side; of the *latter,* fibres pass into the gray matter, to the posterior vesicular columns, passing obliquely through the posterior white columns upward and downward for some distance, and enter the gray matter at different heights.

FIG. 11.

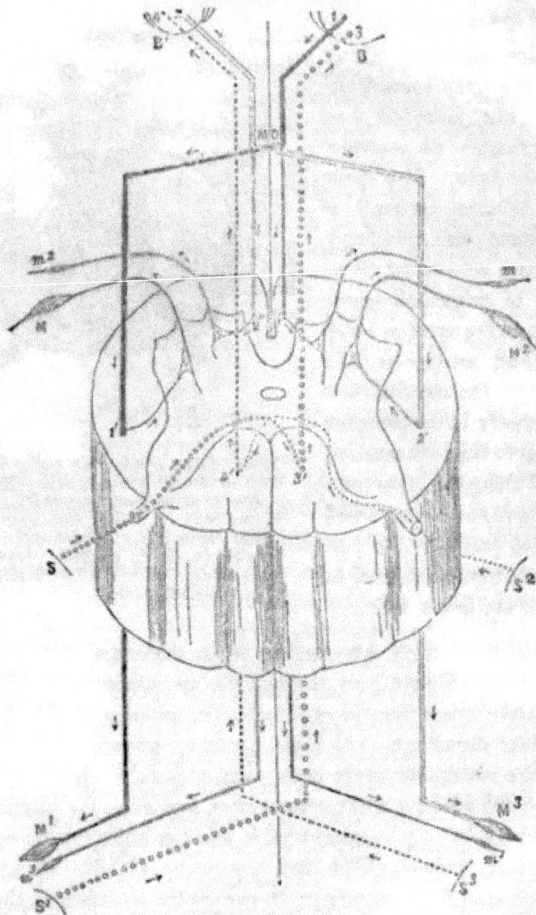

Diagram showing the course, through the spinal cord, of the motor and sensory nerve fibres. B and B' represent the right and left hemispheres of the brain, from which the motor fibres take their origin, and in which the sensory fibres terminate. The *motor tract* from the right side [1] passes down through the crus, through the pons to the medulla oblongata, where it divides into two portions; 1st, the *larger portion*, ninety-seven per cent., crosses over to the opposite side of the cord and passes down through the lateral column. It gives off fibres at different levels, which pass into the gray matter and become connected with the muscles, M, through the multipolar cells; the *smaller portion*, three per cent., does not cross over, but descends on the same side of the cord in the anterior column and supplies the muscles, *m*. The same is true for the motor tract for the left hemisphere.

The *sensory fibres* from the left side of the body enter the gray matter through the posterior roots. They then cross over at once to the opposite side of the cord and ascend to the hemisphere partly in the gray matter, partly in the posterior column. The same is true for the sensory nerves of the right side of the body.

Decussation of Motor and Sensory Fibres. The *Motor* fibres, which conduct volitional impulses from the brain outward to the anterior cornuæ, arise in the motor centres of the cerebrum; they then pass downward through the corona radiata, the internal capsule, the inferior portions of the crura cerebri, the pons Varolii, to the medulla oblongata, where the motor tract of each side divides into two portions, viz. : **1.** The *larger*, containing 91 to 97 per cent. of the fibres, which decussates at the lower border of the medulla and passes down in the *lateral column* of the opposite side, and constitutes the *crossed pyramidal tract*. 2. The *smaller*, containing 3 to 9 per cent. of the fibres, does not decussate, but passes down the *anterior column* of the same side, and constitutes the *direct pyramidal tract*, or the column of Türck. Some of the motor fibres of these two tracts, after entering the anterior cornuæ of the gray matter, become connected with the large multipolar nerve cells, while others pass directly into the anterior roots. Through this decussation each half of the brain governs the muscular movements of the opposite side of the body.

The *Sensory* fibres, which convey the impression made upon the periphery to the cord and brain, pass into the cord through the posterior roots of spinal nerves ; they then diverge and enter the gray matter at different levels, and at once decussate, passing to the opposite side of the gray matter. The *sensory tract* passes upward, through the cord, the medulla, pons Varolii, the superior portion of the crura cerebri, the posterior third of the internal capsule, to the sensory perceptive centre, located in the hippocampus major and unciate convolution (Ferrier). Through this decussation each half of the brain governs the sensibility of the opposite half of the body.

Properties of the Spinal Cord. Irritation applied directly to the *antero-lateral* white columns produces muscular movements but no pain ; they are, therefore, *excitable* but *insensible*.

The surface of the *posterior* columns is very sensitive to direct irritation, especially near the origin of the posterior roots; less so toward the posterior median fissure. The sensibility is due, however, *not* to its own proper fibres, but to the fibres of the posterior roots which traverse it.

Division of the *antero-lateral* columns abolishes all power of voluntary movement in the lower extremities.

Division of the *posterior* columns impairs the power of muscular coördination, such as is witnessed in locomotor ataxia.

The gray matter is probably both *insensible* and *inexcitable* under the influence of direct stimulation.

A *transverse section* of one lateral half of the cord produces :—

(1) On the *same* side, paralysis of voluntary motion and a relative or absolute elevation of temperature and an increased flow of blood in the paralyzed parts; hyperæsthesia for the sense of contact, tickling, pain and temperature.

(2) On the *opposite* side, complete anæsthesia as regards contact, and tickling and temperature, in the parts corresponding to those which are paralyzed in the opposite side. Complete preservation of voluntary power and of the muscular sense.

A *vertical* section through the middle of the *gray matter* results in the loss of sensation on both sides of the body below the section, but no loss of voluntary power.

FUNCTIONS OF THE SPINAL CORD.

1. **As a Conductor.** The *Lateral columns*, particularly the posterior portions, the " pyramidal tracts," and the columns of Türck, are the channels through which pass the voluntary motor impulses from the brain to the large multipolar nerve cells in the anterior cornuæ of gray matter, and through them become connected with the anterior roots which transmit the motor stimuli to the muscles.

The *Anterior columns*, especially the portion surrounding the anterior cornuæ, the " anterior radicular zones," are composed of short longitudinal commissural fibres, which serve to connect together different segments of the spinal cord, a condition required for the coördination of muscular movements.

The *Posterior columns* are composed of short and long commissural fibres which connect together different segments of the cord. They are insensible to direct irritation, but aid in the coördination of muscular movements in walking, standing, running, etc. Degeneration of the posterior columns give rise to the lack of muscular coördination observed in locomotor ataxia.

The *Gray matter*, and especially that portion immediately surrounding the central canal, transmits the sensory nerve fibres from the posterior roots up to the brain. Decussation of the sensory fibres takes place throughout the whole length of the gray matter.

The *Multipolar cells* of the *anterior cornuæ* are connected with the generation and transmission of motor impulses outward; are centres for reflex movements; are the trophic centres for the motor nerves and muscular fibres to which they are distributed. The anterior roots give passage to the vaso-constrictor and vaso-dilator fibres which exert an influence upon the calibre of the blood vessels. *Complete destruction* of the anterior horns is followed by a paralysis of motion, degeneration of the anterior roots, atrophy of muscles and bones and an abolition of reflex movements.

2. As an Independent Nerve Centre.

The spinal cord, by virtue of its containing ganglionic nerve matter, is capable of transforming impressions made upon the centripetal nerves into motor impulses, which are reflected outward through centrifugal nerves to muscles, producing movements. These reflex movements taking place through the gray matter, are independent of sensation and volition.

The mechanism involved in every reflex act is a sentient surface, a sensory nerve, a nerve centre, a motor nerve and muscle.

The *reflex excitability* of the cord may be—

(1) *Increased* by disease of the lateral columns, the administration of strychnia, and in frogs, by a separation of the cord from the brain, the latter apparently exerting an inhibitory influence over the former and depressing its reflex activity.

2. *Inhibited* by destructive lesions of the cord, *e. g.*, locomotor ataxia, atrophy of the anterior cornuæ, the administration of various drugs, and, in the frog, by irritation of certain regions of the brain. When the cerebrum alone is removed and the optic lobes stimulated, the time elapsing between the application of an irritant to a sensory surface and the resulting movement will be considerably prolonged. The optic lobes (Setchenow's centre) apparently generating impulses which, descending the cord, retard its reflex movements.

All movements taking place through the nervous system are of this reflex character, and may be divided into *excito-motor, sensori-motor*, and *ideo-motor*.

Classification of Reflex Movements. (Küss.) They may be divided into four groups, according to the route through which the centripetal and centrifugal impulses pass.

1. Those normal reflex acts, *e. g.*, deglutition, coughing, sneezing, walking, etc., pathological reflex acts, *e. g.*, tetanus, vomiting, epilepsy, which take place both centripetally and centrifugally, through spinal nerves.

2. Reflex acts which take place in a centripetal direction through a cerebro-spinal sensory nerve, and in a centrifugal direction through a sympathetic motor nerve, usually a vasomotor nerve, *e. g.*, the normal reflex acts, which give rise to most of the secretions, pallor and blushing of the skin, certain movements of the iris, certain modifications in the beat of the heart ; the pathological, which, on account of the difficulty in explaining their production, are termed *metastatic, e. g.*, ophthalmia, coryza, orchitis,

which depend on a reflex hyperæmia; amaurosis, paralysis, paraplegia, etc., due to a reflex anæmia.

3. Reflex movements, in which the centripetal impulse passes through a sympathetic nerve, and the centrifugal through a cerebro-spinal nerve; most of these phenomena are pathological, *e. g.*, convulsions from intestinal irritation produced by the presence of worms, eclampsia, hysteria, etc.

4. Reflex actions, in which both the centripetal and centrifugal impulses pass through filaments of the sympathetic nervous system, *e. g.*, those obscure reflex actions which preside over the secretions of the intestinal fluids, which unite the phenomena of the generative organs, the dilatation of the pupils **from** intestinal irritation (worms), and many pathological phenomena.

Laws of Reflex Action. (Pflüger.)

1. *Law of Unilaterality.* If a feeble irritation be applied **to one or** more sensory nerves, movement takes place usually on one side only, and that upon the same side as the irritation.

2. *Law of Symmetry.* If the irritation becomes sufficiently intense, motor reaction is manifested, in addition, in corresponding muscles of the opposite side of the body.

3. *Law of Intensity.* Reflex movements are usually more intense on the side of the irritation; at times the movements of the opposite side equal them in intensity, but they are usually less pronounced.

4. *Law of Radiation.* If the excitation still continues to increase, it is propagated upward, and motor reaction takes place through centrifugal nerves coming from segments of the cord higher up.

5. *Law of Generalization.* When the irritation becomes very intense, it is propagated to the medulla oblongata; motor reaction then becomes general, and is propagated up and down the cord, so that all the muscles of the body are thrown into action, the medulla oblongata acting as a focus **whence** radiate all reflex **movements.**

Special Centres in the Spinal Cord.

Genito-spinal centre. In the lower portion of the spinal cord are located the centres which control the sphincter muscles of the rectum and bladder, the erection of the penis, the emission of the semen, **the** action of the uterus during parturition, etc.

Cilio-spinal centre. Situated in **the** spinal cord between the 6th cervical and 2d dorsal nerves; stimulation of the cord in this situation produces a dilatation of both pupils through filaments of the sympathetic, which take their origin from this region of the cord.

Throughout the spinal cord are situated numerous centres which preside over the following reflexes, viz. :—

The *patellar tendon reflex* takes place through the segments from which arise the 2d, 3d and 4th lumbar nerves ; the *cremasteric reflex* through the segment from which arise the 1st and 2d lumbar nerves ; the *abdominal reflex* through the segments between the 8th and 12th dorsal nerves ; the *epigastric reflex* through the segments from which arise the 4th, 5th and 6th dorsal nerves.

Paralysis from Disease of the Spinal Cord.

Seat of Lesion. If it be in the *lower part* of the *sacral canal*, there is paralysis **of the** compressor urethræ, accelerator urinæ, and sphincter ani muscles ; **no paralysis of the** muscles of the leg.

At the upper limit of the sacral region. Paralysis of the muscles of the bladder, rectum and anus ; loss of sensation and motion in the muscles of the legs, except those supplied by the anterior crural and **obturator,** viz. : psoas iliacus, Sartorius, pectineus, adductor longus, magnus **and** brevis, obturator, vastus externus and internus, etc.

At the upper limit of the lumbar region. Sensation and **motion** paralyzed in both legs ; loss of power over the rectum and **bladder ; paralysis** of the muscular walls of the abdomen interfering with **expiratory movements.**

At the lower portion of the cervical region. Paralysis **of the legs, etc.,** as above ; in addition, paralysis of all the intercostal **muscles and consequent** interference with respiratory movements ; paralysis **of muscles of the upper extremities,** except those of the shoulders.

Above the middle of the cervical region. In addition to the preceding, difficulty of deglutition **and** vocalization, contraction of the pupils, paralysis of the diaphragm, scalene muscles, intercostals, and many of the accessory respiratory muscles ; death resulting immediately, from arrest of respiratory movements.

Anterior half of spinal cord. Paraplegia developing symmetrically.

Posterior half of spinal cord. Characteristic symptoms of locomotor ataxia or tabes dorsalis.

In the gray substance in the vicinity of the central canal and anterior horns. If the lesion be acute, symptoms characteristic of acute spinal paralysis manifest themselves ; if chronic, symptoms characteristic of progressive muscular atrophy.

MEDULLA OBLONGATA.

The Medulla Oblongata is the expanded portion of the upper part of
the spinal cord.   It is pyramidal in form and measures one and a half
inches in length, three-quarters of an inch in breadth, half an inch in
thickness, and is divided into two lateral halves by the anterior and posterior median fissures, which are continuous with those of the cord.   Each
half is again subdivided by minor grooves, into four columns, viz. : *anterior*

FIG. 12.

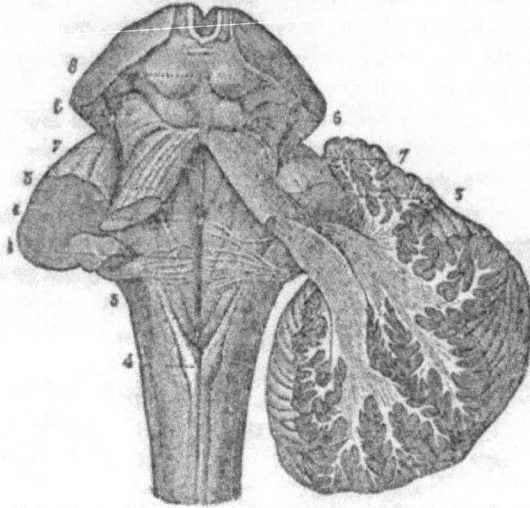

View of Cerebellum in section, and of Fourth Ventricle, with the neighboring parts.
(*From Sappey.*)

1. Median groove fourth ventricle, ending below in the *calamus scriptorius,* with the
longitudinal eminences formed by the fasciculi teretes, one on each side.   2. The same
groove, at the place where the white streaks of the auditory nerve emerge from it to
cross the floor of the ventricle.   3. Inferior peduncle of the cerebellum, formed by the
restiform body.   4. Posterior pyramid, above this is the calamus scriptorius.   5 Superior peduncle of cerebellum, or *processus e cerebello ad testes.*   66. Fillet to the side
of the crura cerebri.   7 7. Lateral grooves of the crura cerebri.   8. *Corpora quadrigemina.—After Hirschfeld and Leveille.*

pyramid, lateral tract and *olivary body, restiform body* and *posterior
pyramid.*

1. The *anterior pyramid* is composed partly of fibres continuous with
those of the anterior column of the spinal cord; but mainly of fibres derived from the lateral tract of the opposite side, by *decussation.*   The
united fibres then pass upward through the pons Varolii and crura cerebri,
and for the most part terminate in the corpus striatum and cerebrum.

2. The *lateral tract* is continuous with the lateral columns of the cord;

its fibres in passing upward take three directions, viz.: an external bundle joins the restiform body, and passes into the cerebellum; an internal bundle decussates at the median line and joins the opposite anterior pyramid; a middle bundle ascends beneath the olivary body, behind the pons, to the cerebrum, as the *fasciculus teres*.

The *olivary body* of each side is an oval mass, situated between the anterior pyramid and restiform body; it is composed of white matter externally and gray matter internally, forming the *corpus dentatum*.

3. The *restiform body*, continuous with the posterior column of the cord, also receives fibres from the lateral column. As the restiform bodies pass upward they diverge and form a space, the 4th ventricle, the floor of which is formed by gray matter, and then turn backward and enter the cerebellum.

4. The *posterior pyramid* is a narrow, white cord bordering the posterior median fissure; it is continued upward, in connection with the *fasciculus teres*, to the cerebrum.

The Gray Matter of the medulla is continuous with that of the cord. It is arranged with much less regularity, becoming blended with the white matter of the different columns, with the exception of the anterior. By the separation of the posterior columns, the transverse commissure is exposed, forming part of the floor of the 4th ventricle; special collections of gray matter are found in the posterior portions of the medulla, connected with the roots of origin of different cranial nerves.

Properties and Functions. The medulla is excitable anteriorly, and sensitive posteriorly to direct irritation. It serves (1) as a conductor of sensitive impressions upward from the cord, through the gray matter to the cerebrum; (2) as a conductor of voluntary impulses from the brain to the spinal cord and nerves, through its anterior pyramids; (3) as a conductor of coördinating impulses from the cerebellum, through the restiform bodies to the spinal cord.

As an Independent Reflex Centre. The medulla oblongata contains special collections of gray matter, which constitute independent nerve centres which preside over different functions, some of which are as follows, viz. :—

1. A centre which controls the movements of *mastication*, through afferent and efferent nerves. (See page 24.)

2. A centre reflecting impressions which influence the *secretion* of *saliva*. (See page 26.)

3. A *centre for sucking, mastication and deglutition*, whence are derived

motor stimuli exciting to action and coördinating the muscles of the palate, pharynx and œsophagus, necessary for the swallowing of the food. *The secretion of saliva* is also controlled by a centre in the medulla.

NERVOUS CIRCLE OF DEGLUTITION. (2d and 3d Stages.)

Excitor or Centripetal Nerves.	Palatal branch of 5th pair. Pharyngeal branches of the glosso-pharyngeal. Superior laryngeal branches of the pneumogastric. Œsophageal branches of the pneumogastric.
Motor or Centrifugal Nerves.	Pharyngeal branches of the pneumogastric, derived from the spinal accessory. Hypoglossal and branches of the cervical plexus. Inferior or recurrent laryngeal. Motor filaments of the 3d division of the 5th pair. Portio dura.

4. A centre which coördinates the muscles concerned in the act of *vomiting.*

5. A *Speech centre*, coördinating the various muscles necessary for the accomplishment of articulation through the hypoglossal, facial nerves and the 2d division of the 5th pair.

6. A centre for the harmonization of muscles concerned in *expression,* reflecting its impulses through the facial nerve.

7. A *Cardiac centre,* which exerts (1) an *accelerating* influence over the heart's pulsations through accelerating nerve fibres emerging from the cervical portion of the cord, entering the inferior cervical ganglion, and thence passing to the heart; (2) an *inhibitory* or retarding influence upon the action of the heart, through fibres of the spinal accessory nerve running in the trunk of the pneumogastric.

8. A *Vasomotor centre,* which, by alternately contracting and dilating the blood vessels through nerves distributed in their walls, regulates the quantity of blood distributed to an organ or tissue, and thus influences nutrition, secretion and calorification. The vasomotor centre is situated in the medulla oblongata and pons Varolii, between the *corpora quadrigemina* and the *calamus scriptorius.* The vasomotor fibres having their origin in this centre descend through the interior of the cord, emerge through the anterior roots of spinal nerves, enter the ganglia of the sympathetic, and thence pass to the walls of the blood vessels, and maintain the *arterial tonus*; they may be divided into two classes, viz.: *vaso-dilators, e. g.,* chorda tympani, and *vaso-constrictors, e. g.,* sympathetic fibres.

Division of the cord at the lower border of the medulla is followed by a dilatation of the entire vascular system and a marked fall of the blood pressure. *Galvanic stimulation* of the divided surface of the cord is followed by a contraction of the blood vessels and a rise in the blood pressure.

9. A *Diabetic centre*, irritation of which causes an increase in the amount of urine secreted, and the appearance of a considerable quantity of sugar.

10. *Respiratory centre*, situated near the origin of the pneumogastric nerves, presides over the movements of respiration and its modifications, laughing, sighing, sobbing, sneezing, etc. It may be excited *reflexly* by the presence of carbonic acid in the lungs irritating the terminal pneumo-gastric filaments; or *automatically*, according to the character of the blood circulating through it; an *excess* of carbonic acid or a diminution of oxygen increasing the number of respiratory movements; a reverse condition di-minishing the respiratory movements.

11. A *Spasm centre*, stimulation of which gives rise to convulsive phe-nomena, such as coughing, sneezing, etc.

12. A *centre for certain ocular functions*, governing the closure of the eyelids and dilatation of the pupil.

13. A *Sweat centre* is also localized in the medulla.

NERVOUS CIRCLE OF RESPIRATION (ENTIRELY REFLEX).

Excitor or Centripetal Nerves.	Pulmonary branches of the pneumogastric. Superior laryngeal. Trifacial, or 5th pair. Nerves of general sensibility. Sympathetic nerve.
Motor or Centrifugal Nerves.	Phrenic, distributed to the diaphragm. Intercostals, distributed to the intercostal muscles. Facial nerve, or portio dura, to the facial muscles. External branch of spinal accessory, to the trapezius and sterno-cleido-mastoid muscles.

PONS VAROLII.

The Pons Varolii unites together the cerebrum above, the cerebellum behind, and the medulla oblongata below. It consists of transverse and longitudinal fibres, amidst which are irregularly scattered collections of gray or vesicular nervous matter.

The *transverse fibres* unite the two lateral halves of the cerebellum.

The *longitudinal fibres* are continuous (1) with the anterior pyramids of the medulla oblongata, which interlacing with the deep layers of the transverse fibres, ascend to the crura cerebri, forming their superficial or fasciculated portions; (2) with fibres derived from the olivary fasciculus,

some of which pass to the tubercula quadrigemina, while others, uniting
with fibres from the lateral and posterior columns of the medulla, ascend
in the deep or posterior portions of the crura cerebri.

Properties and Functions. The superficial portion is *insensible* and
inexcitable to direct irritation; the deeper portions appear to be *excitable*,
consisting of descending motor fibres; the posterior portions are *sensible*
but *inexcitable* to irritation.

Transmits motor impulses and sensory impressions from and to the
cerebrum.

The *gray ganglionic* matter consists of centres which convert impressions into conscious sensations, and originate motor impulses, these taking
place independent of any intellectual process; they are the seat of instinctive reflex acts; the centres which assist in the coördination of the automatic movements of station and progression.

CRURA CEREBRI.

The Crura Cerebri are largely composed of the longitudinal fibres of
the pons (anterior pyramids, fasciculi teretes); after emerging from the
pons they increase in size, and become separated into two portions by a
layer of dark gray matter, the *locus niger*.

The *superficial* portion, the *crusta*, composed of the anterior pyramids,
constitutes the *motor tract*, which terminates, for the most part, in the
corpus striatum, but to some extent, also, in the cerebrum; the *deep portion*, made up of the fasciculi teretes and posterior pyramids and accessory
fibres from the cerebellum, constitute the *sensory tract* (the *tegmentum*),
which terminates in the *optic thalamus* and cerebrum.

Function. The crura are conductors of motor impulses and sensory
impressions; the gray matter, the *locus niger*, assists in the coördination of
the complicated movements of the eyeball and iris, through the motor oculi
communis nerve. They also assist in the harmonization of general muscular movements; section of one crus giving rise to peculiar movements of
rotation and somersaults forward and backward.

CORPORA QUADRIGEMINA.

The Corpora Quadrigemina are four small, rounded eminences, two
on each side of the median line, situated immediately behind the third
ventricle, and beneath the posterior border of the corpus callosum.

The *anterior* tubercles are oblong from before backward, and larger than the *posterior*, which are hemispherical in shape; they are grayish in color, but consist of white matter externally and gray matter internally.

Both the anterior and posterior tubercles are connected with the optic thalami by commissural bands named the *anterior* and *posterior brachia*, respectively. They receive fibres from the olivary fasciculus and fibres from the cerebellum, which pass upward to enter the optic thalami.

The *corpora geniculata* are situated, one on the inner side and one on the outer side of each optic tract, behind and beneath the optic thalamus, and from their position are named the *corpora geniculata interna* and *externa*; they give origin to fibres of the optic nerve.

Functions. The *Tubercula quadrigemina* are the physical centres of sight, translating the luminous impressions into visual sensations. Destruction of these tubercles is immediately followed by a loss of the sense of sight; moreover, their action in vision is crossed, owing to the decussation of the optic tracts, so that if the tubercle of the right side be destroyed by disease or extirpated, in a pigeon, the sight is *lost* in the eye of the opposite side, and the iris loses its mobility.

The tubercula quadrigemina as nerve centres preside over the reflex movements which cause a dilation or contraction of the iris; irritation of the tubercles causing contraction, destruction causing dilatation. Removal of the tubercles on one side produces a temporary loss of power of the opposite side of the body, and a tendency to move around an axis is manifested, as after a section of one crus cerebri, which, however, may be due to giddiness and loss of sight.

They also assist in the coördination of the complex movements of the eye, and regulate the movements of the iris during the movements of accommodation for distance.

CORPORA STRIATA AND OPTIC THALAMI.

The Corpora Striata are two large ovoid collections of gray matter, situated at the base of the cerebrum, the larger portions of which are imbedded in the white matter, the smaller portions projecting into the anterior part of the lateral ventricle. Each striated body is divided, by a narrow band of white matter, into two portions, viz.:—

1. The *Caudate nucleus*, the intraventricular portion, which is conical in shape, having its apex directed backward, as a narrow, tail-like process.

2. The *Lenticular nucleus*, imbedded in the white matter, and for the most part external to the ventricle; on the outer side of the lenticular nucleus is found a narrow band of white matter, the *external capsule*;

and between it and the convolutions of the island of Reil, a thin band of gray matter, the *claustrum;* the **corpora striata are** grayish in color, and when divided present transverse striations, from the intermingling of white fibres and gray cells.

The **Optic Thalami** are two oblong masses situated in the ventricles posterior to the corpora striata, and resting upon the posterior portion of the crura cerebri. The internal surface projecting into the lateral ventricles is white, but the interior is grayish, from a commingling of both white fibres **and gray** cells. Separating the lenticular nucleus from **the caudate nucleus** and the optic thalamus, is a band of white tissue, the *internal capsule.*

The *internal capsule* is a narrow, bent tract of white matter, and is, for the most part, an expansion of the *motor tract* of the crura cerebri. It **consists of** two segments, *an anterior,* situated between **the** caudate nucleus and the anterior surface of the lenticular nucleus, and a *posterior,* situated between the optic thalamus and the posterior surface of the lenticular nucleus. These two segments unite at an obtuse angle, which is directed toward the median line. Pathological observation has shown that the nerve fibres of the direct and crossed pyramidal tracts can be traced upward through the anterior two-thirds of the posterior segment, into the centrum ovale, where, for the most part, **they are** lost; a portion, however, remaining united, ascend higher and terminate in the paracentral lobule, the superior extremity of the ascending frontal and parietal convolutions. The *sensory tract* can be **traced** upward, through the posterior third, into the cerebrum, where they probably terminate in the hippocampus major and unciate **convolution.**

Functions. The *Corpora striata* **are** the centres in which terminate some **of the fibres of the superficial or** *motor* **tract** of the crura cerebri ; others pass upward through the *internal capsule,* to be distributed to the cerebrum. It might be inferred, from their anatomical relations, that they are motor centres. *Irritation* by a weak galvanic current produces muscular movements of the opposite side of the body; *destruction* of their substance by a hemorrhage, as in apoplexy, is followed by a paralysis of motion **of** the opposite side of the body, but there is no loss of sensation. When the hemorrhagic destruction involves the fibres of the anterior two-thirds of the posterior segment of the internal capsule, and thus separates them from their trophic centres in the cortical motor region, a descending degeneration is established, which involves the direct pyramidal tract of the same side and the crossed pyramidal tract of the opposite side.

Destruction of the posterior one-third of the posterior segment of the internal capsule is followed by a loss of sensation on the opposite side of the body, and a loss of the senses of smell and vision on the same side (Charcot). The precise function of the corpora striata is unknown, but they are in some way connected with motion.

The *Optic thalami* receives the fibres of the *tegmentum*, the posterior portion of the crura cerebri. They are insensible and inexcitable to direct irritation. Removal of one optic thalamus, or destruction of its substance by disease or hemorrhage, is followed by a loss of sensibility of the opposite side of the body, but there is no loss of motion; their precise function is also unknown, but in some way connected with *sensation*. In both cases their action is crossed.

CEREBELLUM.

The Cerebellum is situated in the inferior fossæ of the occipital bone, beneath the posterior lobes of the cerebrum. It attains its maximum weight, which is about 5 ozs., between the twenty-fifth and fortieth years; the proportion between the cerebellum and cerebrum being 1 to 8½.

It is composed of *two lateral hemispheres* and a central elongated lobe, the *vermiform process;* the two hemispheres are connected with each other by the fibres of the *middle peduncle* forming the superficial portion of the pons Varolii. It is brought into connection with the *medulla oblongata* and *spinal cord*, through the prolongation of the *restiform* bodies; with the cerebrum, by fibres passing upward beneath the corpora quadrigemina and the optic thalami, and then forming part of the diverging cerebral fibres.

Structure. It is composed of both white and gray matter, the former being internal, the latter external, and convoluted, for economy of space.

The *White matter* consists of a central stem, the interior of which is a dentated capsule of gray matter, the *corpus dentatum*. From the external surface of the stem of white matter processes are given off, forming the *laminæ*, which are covered with gray matter.

The *Gray matter* is convoluted and covers externally the laminated processes; a vertical section through the gray matter reveals the following structures :—

1. A delicate *connective tissue layer*, just beneath the pia mater, containing rounded corpuscles, and branching fibres passing toward the external surface.

2. *The cells of Purkinje*, forming a layer of large, nucleated, branched nerve cells sending off processes to the external layer.   ·

I

3. *A granular layer* of small, but numerous corpuscles.

4. *Nerve fibre layer*, formed by a portion of the white matter.

Properties and Functions. Irritation of the cerebellum is not followed by any evidences either of pain or convulsive movements; it is, therefore, *insensible* and *inexcitable*.

Co-ordination of Movements. Removal of the superficial portions of the cerebellum in pigeons produces *feebleness* and *want of harmony* in the muscular **movements;** as successive slices are removed, the movements become more **irregular, and** the pigeon becomes restless; when the last portions **are removed, all power** of *flying, walking, standing*, etc., is entirely gone, and the *equilibrium* cannot be maintained, the power of *coördinating* muscular movements being entirely gone. **The same** results **have been** obtained by operating on all classes of animals.

The following symptoms were noticed by Wagner, after **removing the** whole or a large part of the cerebellum. 1. A tendency on the part **of** the animal to throw itself on one side, and to extend the legs as far as possible. 2. Torsion of the head on the neck. 3. Trembling of the muscles of the body, which was general. 4. Vomiting and occasionally liquid evacuations.

Forced Movements. Division of *one crus cerebelli* causes the animal to fall on one side and roll rapidly on its longitudinal axis. According to Schiff, if the peduncle be divided from *behind*, the animal falls on the same side as the injury; if the section be made in *front*, the animal **turns** to the opposite side.

Disease **of the** cerebellum partially corroborates the **result** of experiments; in many cases symptoms of unsteadiness of gait, from *a want of coördination*, have been noticed.

Comparative anatomy reveals a remarkable correspondence between the development of the cerebellum and the complexity of muscular actions. It attains a much greater development, relatively to the rest of the brain, in those animals whose movements are very complex and varied in character, such as the kangaroo, shark and swallow.

The cerebellum may possibly exert some influence over the **sexual function**, but physiological and pathological facts **are** opposed to the **idea of** its being the seat of the sexual instinct. It appears to be simply a **centre** for the coördination and equilibration of muscular movements.

CEREBRUM.

The Cerebrum is the largest portion of the encephalic mass, constituting about four-fifths of its weight; the average weight in the adult male is from 48 to 50 ozs., or about three pounds, while in the adult female it is about five ozs. less. After the age of forty the weight of the cerebrum gradually diminishes at the rate of one ounce every ten years. In idiots the brain weight is often below the normal, at times not amounting to more than twenty ounces.

The **Blood Supply** to the cerebrum is unusually large, considering its comparative bulk; nearly *one-fifth* of the entire volume of blood being distributed to it by the carotid and vertebral arteries. These vessels anastomose so freely, and are so arranged within the cavity of the cranium, that an obstruction in one vessel will not interfere with the regular supply of blood to the parts to which its branches are distributed. A diminished amount, or complete cessation, of the supply of blood is at once followed by a suspension of its functional activity.

The cerebrum is connected with the pons Varolii and medulla oblongata through the crura cerebri, and with the cerebellum, through the superior peduncles. It is divided into two lateral halves, or hemispheres, by the longitudinal fissure running from before backward in the median line; each hemisphere is composed of both *white* and *gray* matter, the former being internal, the latter external; it covers the surfaces of the hemisphere which are infolded, forming convolutions, for economy of space.

Fissures.

1. The *Fissure of Sylvius* is one of the most important; it is the first to appear in the development of the fœtal brain, being visible at about the third month; in the adult it is quite deep and well marked, running from the under surface of the brain upward, outward and backward, and forms a boundary between the frontal and temporo-sphenoidal lobes.

2. The *Fissure of Rolando* is second in importance, and runs from a point on the convexity near the median line transversely outward and downward toward the fissure of Sylvius, but does not enter it. It separates the frontal from the parietal lobe.

3. The *Parietal fissure*, arising a short distance behind the fissure of Rolando, upon the convexity of the hemisphere, runs downward and backward to its posterior extremity.

Secondary fissures of importance are found in different lobes of the cerebrum, separating the various convolutions. In the anterior lobe are found the *pre-central, superior frontal* and *inferior frontal* fissures;

in the occipital lobe are found the *parieto occipital* and the *calcarine fissures.*

Convolutions. *Frontal lobe.*

The *Ascending frontal convolution*, situated in front of the fissure of Rolando, runs downward and forward; it is continuous above with the anterior frontal, and below with the inferior frontal convolution.

FIG. 13.

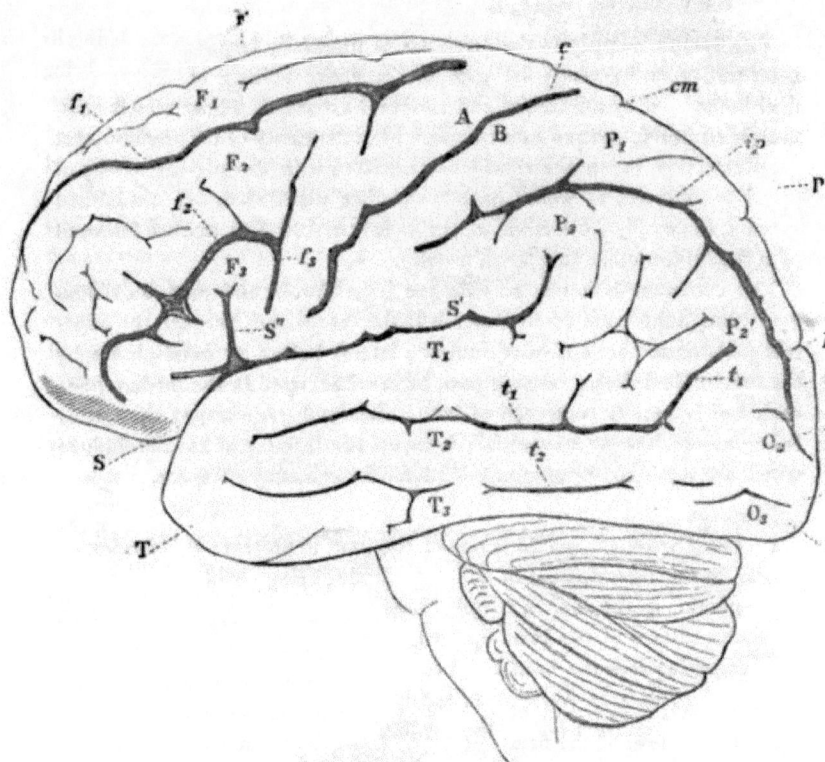

LEFT SIDE OF THE HUMAN BRAIN (DIAGRAMMATIC).

F, frontal; P, parietal; O, occipital; T, temporo-sphenoidal lobe; S, fissure of Sylvius; S', horizontal; S", ascending ramus of S; c, sulcus centralis, or fissure of Rolando; A, ascending frontal, and B, ascending parietal, convolution; F₁, superior, F₂, middle, and F₃, inferior, frontal convolutions; f₁, superior, and f₂, inferior, frontal fissures; f₃, sulcus præcentralis; P, superior parietal lobule; P₂, inferior parietal lobule, consisting of P₂, supra-marginal gyrus, and P'₂, angular gyrus; i p, sulcus interparietalis; c m, termination of calloso-marginal fissure; O₁, first, O₂, second, O₃, third, occipital convolutions; p o, parieto-occipital fissure; o, transverse occipital fissure; o₃, inferior longitudinal occipital fissure; T₁, first, T₂, second, T₃, temporo-sphenoidal, convolutions, t₁, first, t₂, second, temporo-sphenoidal fissures. (*Landois' Physiology.*)

The *Superior frontal convolution* is bounded internally by the longitudinal fissure, and externally by the superior frontal fissure ; it is connected with the superior end of the frontal convolution, and runs downward and forward to the anterior extremity of the frontal lobe, where it turns backward, and rests upon the orbital plate of the frontal bone.

The *Middle frontal convolution*, the largest of the three, runs from behind forward, along the sides of the lobe, to its anterior part ; it is bounded above by the superior and below by the inferior frontal fissures.

The *Inferior frontal convolution* winds around the ascending branch of the fissure of Sylvius, in the anterior and inferior portion of the cerebrum.

Parietal Lobe. The *Ascending parietal convolution* is situated just behind the fissure of Rolando, running downward and forward ; above, it becomes continuous with the upper parietal convolution, and below, winds around to be united with the ascending frontal.

The *Upper parietal convolution* is situated between the parietal and longitudinal fissures.

The *Supra-marginal convolution* winds around the superior extremity of the fissure of Sylvius.

The *Angular convolution*, a continuation of the preceding, follows the parietal fissure to its posterior extremity, and then makes a sharp angle downward and forward.

Temporo-sphenoidal Lobe. Contains three well-marked convolutions, the *superior, middle* and *inferior*, separated by well-defined fissures, and continuous posteriorly with the convolutions of the parietal lobe.

The *Occipital Lobe* lies behind the parieto-occipital fissure, and contains the *superior, middle* and *inferior* convolutions, not well marked.

The *Central Lobe*, or *Island of Reil*, situated at the bifurcation of the fissure of Sylvius, is a triangular shaped cluster of six convolutions, the *gyri operti*, which are connected with those of the frontal, parietal, and temporo-sphenoidal lobes.

Structure. The *Gray matter* of the cerebrum, about one-eighth of an inch thick, is composed of five layers of nerve cells : (1) a superficial layer, containing few small multipolar ganglion cells; (2) small ganglion cells, pyramidal in shape; (3) a layer of large pyramidal ganglion cells with processes running off superiorly and laterally; (4) the granular formation containing nerve cells ; (5) spindle-shaped and branching nerve cells of moderate size.

The *White matter* consists of three distinct sets of fibres :—

1. The *diverging* or *peduncular* fibres are mainly derived from the

columns of the cord and medulla oblongata; passing upward through the crura cerebri, they receive accessory fibres from the olivary fasciculus, corpora quadrigemina and cerebellum. Some of the fibres terminate in the optic thalami and corpora striata, while others radiate into the anterior middle and posterior lobes of the cerebrum.

2. The *transverse commissural* fibres connect together the two hemispheres, through the corpus callosum and anterior and posterior commissures.

3. The *longitudinal commissural* fibres connect together different parts of the same hemisphere.

Functions. The cerebral hemispheres are the centres of the nervous system through which are manifested all the phenomena of the mind; they are the centres in which impressions are registered, and reproduced subsequently as ideas; they are the seat of intelligence, reason and will.

However important a centre the cerebrum may be, for the exhibition of this highest form of nervous action, it is not directly essential for the continuance of life; for it does not exert any control over those automatic reflex acts, such as respiration, circulation, etc., which regulate the functions of organic life.

From the study of comparative anatomy, pathology, vivisection, etc., evidence has been obtained which throws some light upon the physiology of the cerebral hemispheres.

1. *Comparative Anatomy* shows that there is a general connection between the size of the brain, its texture, the depth and number of convolutions, and the exhibition of mental power. Throughout the entire animal series, the increase in intelligence goes hand in hand with an increase in the development of the brain. In man there is an enormous increase in size over that of the highest animals, the anthropoids. The most cultivated races of men have the greatest cranial capacity; that of the educated European being about 116 cubic inches, that of the Australian being about 60 cubic inches, a difference of 56 cubic inches. Men distinguished for great mental power usually have large and well-developed brains; that of Cuvier weighed 64 ozs.; that of Abercrombie 63 ozs.; the average being about 48 to 50 ozs.; not only the size, but above all, the texture of the brain, must be taken into consideration.

2. *Pathology.* Any severe injury or disease disorganizing the hemispheres is at once attended by a disturbance, or entire suspension of mental activity. A blow on the head producing concussion, or undue pressure from cerebral hemorrhage destroys consciousness; physical and chemical alterations in the gray matter have been shown to coexist with insanity,

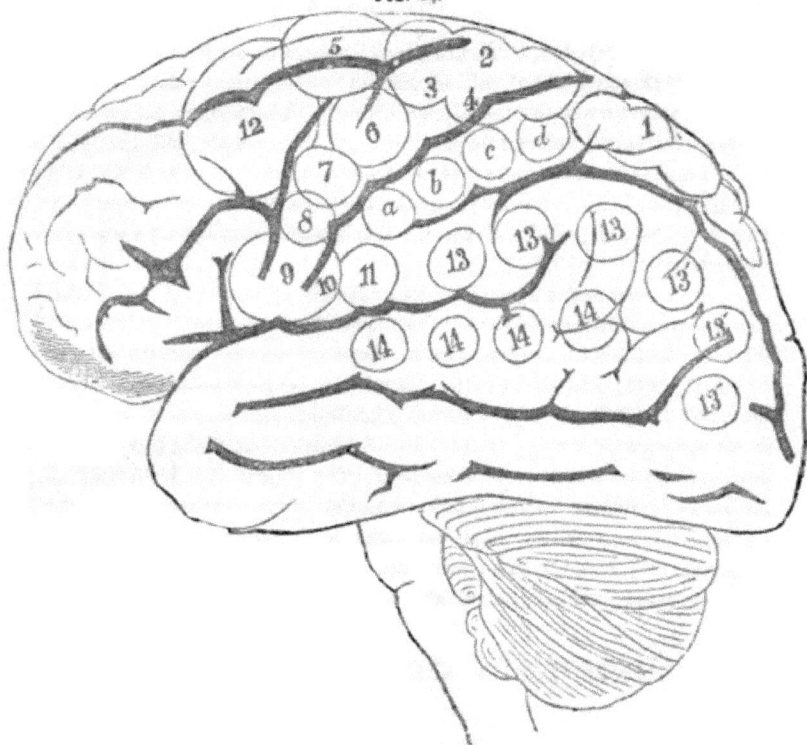

FIG. 14.

SIDE VIEW OF THE BRAIN OF MAN, WITH THE AREAS OF THE CEREBRAL CONVOLUTIONS, ACCORDING TO FERRIER.

The figures are constructed by marking on the brain of man, in their respective situations, the areas of the brain of the monkey as determined by experiment, and the description of the effects of stimulating the various areas refers to the brain of the monkey.

(1) Advance of the opposite hind limb, as in walking.

(2), (3), (4) Complex movements of the opposite leg and arm, and of the trunk, as in swimming.

(a), (b), (c), (d) Individual and combined movements of the fingers and wrist of the opposite hand. Prehensile movements.

(5) Extension forward of the opposite arm and hand.

(6) Supination and flexion of the opposite forearm.

(7) Retraction and elevation of the opposite angle of the mouth, by means of the zygomatic muscles.

(8) Elevation of the ala nasi and upper lip, with depression of the lower lip on the opposite side.

(9), (10) Opening at the mouth, with (9) protrusion and (10) retraction of the tongue; region of aphasia, bilateral action.

(11) Retraction of the opposite angle of the mouth, the head turned slightly to one side.

(12) The eyes open widely, the pupils dilate, and the head and eyes turn toward the opposite side.

(13), (13') The eyes move toward the opposite side with an upward (13) or downward (13') deviation. The pupils are generally contracted.

(14) Pricking of the opposite ear, the head and eyes turn to the opposite side, and the pupils dilate largely.

loss of memory, speech, etc. Congenital defects of organization from imperfect development are usually accompanied by a corresponding deficiency
of intellectual power and the higher instincts. Under these circumstances
no great advance in mental development can be possible, and the intelligence remains at a low grade. In congenital idiocy not only is the brain
of small size, but it is wanting in proper chemical composition; *phosphorus*,
a characteristic ingredient of the nervous tissue, being largely diminished
in amount.

3. *Experimentation* upon the lower animals by removing the cerebral
hemispheres is attended by results similar to those observed in disease and
injury. Removal of the cerebrum in pigeons produces complete abolition
of intelligence, and destroys the capability of performing spontaneous
movements. The pigeon remains in a condition of profound stupor, which
is not accompanied, however, by a loss of sensation, or of the power of producing reflex or instinctive movements. The pigeon can be temporarily
aroused by pinching the feet, loud noises, light placed before the eyes, etc.,
but soon relapses into a state of quietude, being unable to remember impressions and connect them with any train of ideas; the faculties of
memory, reason and judgment being completely abolished.

CEREBRAL LOCALIZATION OF FUNCTION.

From experiments made upon animals, and the results of clinical and
post-mortem observations upon men, it has been shown that the phenomena of organic and psychical life are presided over by anatomically
localized centres in the brain. A knowledge of the position of these centres
becomes of the highest importance in localizing the seat of lesions, thrombi,
hemorrhages, new growths, etc., which show themselves in paralyses,
epilepsies, etc. It has not been possible to thus localize all functions, and to
many parts of the brain no special use can be assigned. The following are the
centres most definitely mapped out and that are of paramount importance :—

Motor Centres. These are in the cortical gray matter, and are arranged along either side of the fissure of Rolando. The upper third of
the ascending frontal and parietal convolutions about this fissure preside
over the movements of the leg of the opposite side of the body ; the middle
third controls the movements of the arm ; the upper part of the inferior
third is the facial area. The lowest part of the inferior third governs the
motility of the lips and tongue, and this space, with the posterior extremity
of the third frontal convolution, constitutes the *speech centre*. The centres
for the ocular muscles lie in the floor of the 4th ventricle. Destruction of

the gray matter at these points is followed by paralysis of the muscles of the opposite side of the body, and morbid growths, hemorrhages or thrombi of the vessels of the parts, result in abnormal stimulation or interference of the functions corresponding to the nature and extent of the lesion. Cerebral or Jacksonian epilepsy is a result of local cortical disease.

Lesions of the third frontal convolution on the left side, if the patient be right handed, produce the various forms of *aphasia* or the partial or complete loss of the power of articulate speech. In deaf mutes this convolution is **very** imperfectly developed.

Aphasia is of many degrees and kinds; *ataxic*, when there is inability to execute the movements of the mouth, etc., necessary to speech; *agraphic*, if there be loss of the power of writing; *amnesic*, when there is loss of memory of words; *paraphasic*, if there is loss of the power to connect rightly the ideas with the proper use of words, etc. In *word blindness* the person cannot name a letter or a word when printed or written; and in *word deafness* the patient hears sounds but does not hear words. Conditions of these kinds are apt to coexist with hemiplegia.

Sensory Centres. These are the centres in which the sensory impressions are coördinated, and in which they probably become parts of our consciousness. The most important are :—

The Visual Centre, located in the occipital lobe and especially in the cuneus. Unilateral destruction of this area results in *hemianopsia*, or blindness of the corresponding halves of the two retinæ. Destruction of both occipital lobes in man results in total blindness. Stimulation or irritation of the visual centre causes *photopsia*, or hallucinations of sight, in corresponding halves of the retinæ. There have been instances of injury of these parts when sensations of color were abolished with preservation of those of space and light, thus showing a special localization of the color centre. Late experiments show that the centres of the two hemipheres are united, as ocular fatigue of a non-used eye was proportional to the fatigue of the exercised one.

The Auditory Centres are located in the temporo-sphenoidal lobes. Word deafness is associated with softening of these parts, and their complete removal results in deafness.

The Gustatory and Olfactory Centres are located in the uncinate gyrus, on the inner side of the temporo-sphenoidal lobes. There does not seem to be any differentiation, up to this time, of these two centres

The *Superior* and *Middle Frontal convolutions* appear to be the seat of the reason, intelligence and will. Destruction of these parts is followed by proportional hebetude, without any impairment of sensation or motion.

SYMPATHETIC NERVOUS SYSTEM.

The Sympathetic Nervous System **consists of a** chain of ganglia connected together by longitudinal nerve filaments, **situated on** each side of the spinal column, running from above downward. **The two** ganglionic cords are connected together in the interior of the cranium by the ganglion of Ribes, on the anterior communicating artery, and terminate in the ganglion impar, situated at the tip of the coccyx.

The chain of ganglia is divided into groups, and named according to the location in which they are found, viz.: *cranial,* four in number; *cervical,* three; *thoracic,* twelve; *lumbar,* five; *sacral,* five; *coccygeal,* one. Each ganglion consists of a collection of *vesicular nervous* matter, among which are found *tubular* and *gelatinous* nerve fibres. The ganglia are reinforced by motor and sensory fibres from the cerebro-spinal nervous system.

The Ganglia are distinct nerve fibres, from which branches are distributed to glands, arteries, muscles, and to the cerebral and spinal nerves; many pass, also, to the visceral ganglia, *e. g.,* cardiac, semilunar, pelvic, etc.

Cephalic Ganglia.

1. The *Ophthalmic* or *Ciliary* ganglion is situated in the orbital cavity posterior **to the** eyeball; it is of small size, and of a reddish-gray color; receives filaments of communication from the motor oculi, ophthalmic branch of the 5th pair, and the carotid plexus. Its filaments of distribution are the ciliary nerves which pass to the iris and ciliary muscle.

Function. It is the centre through which the reflex acts take place by which the pupil is contracted or dilated; controls the movement of accommodation for vision at different distances.

2. The *Spheno-palatine,* or Meckel's ganglion, triangular in shape, is situated in the spheno-maxillary fossa; receives filaments from the facial (Vidian nerve), and the superior maxillary branch of the 5th nerve. Its filaments of distribution pass to the gums, the soft palate, levator palati and azygos uvulæ muscles.

3. The *Otic,* or Arnold's ganglion, is of small size, oval in shape, and situated beneath the foramen ovale; receives a motor filament from the facial and sensory filaments from the glosso-pharyngeal and 5th nerve; sends filaments to the mucous membrane of the tympanic cavity and to the tensor tympani muscle.

4. The *Submaxillary* ganglion, situated in the submaxillary gland, receives motor filaments from the chorda tympani, and sensory filaments from the lingual branch of the 5th nerve. Regulates to some extent the secretion of saliva.

Cervical Ganglia.

The *Superior cervical* ganglion is fusiform in shape, of a grayish-red color, and situate opposite the 2d and 3d cervical vertebræ; it sends branches to form the carotid and cavernous plexuses which follow the course of the carotid arteries to their distribution; also sends branches to join the glosso-pharyngeal and pneumogastric, to **form** the pharyngeal plexus.

The *Middle cervical* ganglion, the smallest of the three, is occasionally wanting; it is situated opposite the 5th cervical vertebra; sends branches to the superior and inferior cervical ganglion, and to the thyroid artery.

The *Inferior cervical* ganglion, irregular in form, is situated opposite the **last** cervical vertebra; it is frequently fused with the first thoracic ganglion.

The *superior, middle* and *inferior cardiac nerves,* arising from these cervical ganglia, pass downward and forward to form the deep and superficial cardiac plexuses located at the bifurcation of the trachea, from which branches are distributed to the heart, coronary arteries, etc.

The **Thoracic Ganglia** are usually twelve in number, **placed** against the heads of the ribs behind the pleura; they are small in size and gray in color; they communicate with the cerebro-spinal nerves by two filaments, one of which is white, the other gray.

The *great splanchnic nerve* is formed by the union of branches from the sixth, seventh, eighth and ninth ganglia; it passes through the diaphragm **to** the semilunar ganglion.

The *lesser splanchnic nerve* is formed by the union of filaments from the tenth and eleventh ganglia, and is distributed to the cœliac plexus.

The *renal splanchnic nerve* arises from the last thoracic ganglion and terminates in the renal plexus.

The *semilunar ganglia,* the largest of the sympathetic, are situated by **the side** of the cœliac axis; they send radiating branches to form the *solar plexus;* from the various plexuses, nerves follow the gastric, splenic, hepatic, renal, etc., arteries, into the different abdominal viscera.

The **Lumbar Ganglia,** four in number, are placed upon the bodies of the vertebra; they give off branches which unite to form the aortic lumbar plexus and the hypogastric plexus, **and** follow the blood vessels to their terminations.

The **Sacral and Coccygeal Ganglia** send filaments of distribution to all the blood vessels of the pelvic viscera.

Properties and Functions. The sympathetic nerve possesses both sensibility and the power of exciting motion, but these properties are much less decided than in the cerebro-spinal system. Irritation of the ganglia does not produce any evidence of pain until some time has elapsed. If caustic soda be applied to the semilunar ganglia, or a galvanic current be passed through the splanchnic nerves, no instantaneous effect is noticed, as in the case of the cerebro-spinal nerves; but in the course of a few seconds a slow, progressive contraction of the muscular coat of the intestines is established, which continues for some time after the irritation is removed. *Division* of the sympathetic nerve in the neck is followed by a vascular congestion of the parts above the section on the corresponding side, attended by an increase in the temperature; not only is there an increase in the amount of blood, but the rapidity of the blood current is very much hastened, and the blood in the veins becomes of a brighter color. *Galvanization* of the upper end of the divided nerve causes all of the preceding phenomena to disappear; the congestion decreases, the temperature falls, and the venous blood becomes dark again.

The sympathetic exerts a similar influence upon the circulation of the limbs and the glandular organs; *destruction* of the first thoracic ganglion and division of the nerves forming the lumbar and sacral plexuses is followed by a dilatation of the vessels, an increased rapidity of the circulation, and an elevation of temperature in the anterior and posterior limbs; galvanization of the peripheral ends of these nerves causes all of these phenomena to disappear. Division of the splanchnic nerve causes a dilatation of the blood vessels of the intestine.

These phenomena of the sympathetic nerve system are dependent upon the presence of vasomotor nerves which, under normal circumstances, exert a tonic influence upon the blood vessels. These nerves, derived from the cerebro-spinal system, the medulla oblongata, leave the spinal cord by the *rami communicantes*, enter the sympathetic ganglia, and finally terminate in the muscular wall of the blood vessels.

Sleep is a periodical condition of the nervous system, in which there is a partial or complete cessation of the activities of the higher nerve centres. The *cause* of sleep is a diminution in the quantity of blood, occasioned by a contraction of the smaller arteries under the influence of the vasomotor nerves.

During the waking state the brain undergoes a physiological waste, as a result of the exercise of its functions; after a certain length of time its activities become enfeebled, and a period of repose ensues, during which a regeneration of its substance takes place.

When the brain becomes enfeebled there is a diminished molecular activity and an accumulation of waste products; under these circumstances it ceases to dominate the medulla oblongata and the spinal cord. These centres then act more vigorously, and diminish the calibre of the cerebral blood vessels through the action of the vasomotor nerves, producing a condition of physiological anæmia and sleep; during this state waste products are removed, force is stored up, nutrition is restored, and waking finally occurs.

THE SENSE OF TOUCH.

The **Sense of** Touch is a modification of general sensibility, **and** located in the skin, which is especially adapted for this purpose, on **account** of **the** number of nerves and papillary elevations it possesses. The **struc**tures of the skin and the modes of termination of the sensory nerves have **already** been considered.

The **Tactile Sensibility** varies in acuteness in different portions of the body; being most marked **in** those regions in which the tactile corpuscles are most abundant, *e. g.*, the palmar surface of the third phalanges of the fingers and thumb.

The *relative sensibility* of different portions of the body has **been** ascertained by means of a pair of compasses, the points of which are guarded by cork, and then determining how closely they could be brought together, and yet be felt at two distinct points. The following are some of the measurements:—

Point of tongue	½ of a line.
Palmar surface of third phalanx	1 line.
Red surface of lips	2 lines.
Palmar surface of metacarpus	3 "
Tip of the nose	3 "
Part of lips covered by skin	4 "
Palm of hand	5 "
Lower part of forehead	10 "
Back of hand	14 "
Dorsum of foot	18 "
Middle of the thigh	30 "

The sense of touch communicates to the mind the idea of *resistance* only, and the varying degrees of resistance offered to the sensory nerves enables us to estimate, with the aid of the muscular sense, the qualities of *hardness* and *softness* of external objects. The idea of *space* or *extension* is obtained when the sensory surface or the external object changes its place in regard to the other, the *character* of the surface, its *roughness* or *smoothness*, is estimated by the impressions made upon the tactile papillæ.

Appreciation of Temperature.—The general surface of the body is more or less sensitive to differences of temperature, though this sensation is separate from that of touch; whether there are **nerves** especially adapted for the conduction of this sensation has not been fully determined. Under pathological conditions, however, the sense of touch may be abolished, while the appreciation of changes in temperature may remain normal.

This cutaneous surface varies in its sensibility to temperature in different parts of the body, and depends, to some extent, upon the thickness of the skin, exposure, habit, etc.; the inner surface of the elbow is more sensitive to changes in temperature than the outer portion of the arm; the left hand is more sensitive than the right; the mucous membrane less so than the skin.

Excessive heat or cold has the same effect upon the sensibility; the temperatures most readily appreciated are those between 50° F. and 115° F.

The sensation of *pain* and *tickling* appear to be conducted to the brain, also, by nerves different from those of touch; in abnormal conditions the appreciation of pain may be entirely lost, while touch remains unimpaired.

THE SENSE OF TASTE.

The Sense of Taste is localized mainly in the mucous membrane covering the superior surface of the tongue.

The Tongue is situated in the floor of the mouth; its base is directed backward, and connected with the hyoid bone, by numerous muscles, with the epiglottis and soft palate; its apex is directed forward against the posterior surface of the teeth.

The *substance* of the tongue is made up of intrinsic muscular fibres, **the** *linguales*; it is attached to surrounding parts, and its various movements performed by the extrinsic muscles, *e. g.*, stylo-glossus, genio-hyoglossus, etc.

The *mucous membrane* covering the tongue is continuous with that lining the commencement of the alimentary canal, and is furnished with vascular and nervous papillæ.

The *papillæ* are analogous in their structure to those of the skin, and are distributed over the dorsum of the tongue, giving it its characteristic roughness.

There are three principal varieties:—

1. The *filiform papillæ* are most numerous, and cover the anterior two-thirds of the tongue; they are conical or filiform in shape, often prolonged into filamentous tufts, of a whitish color, and covered by horny epithelium.

2. The *fungiform papillæ* are found chiefly at the tip and sides of the

tongue; they are larger than the preceding, and may be recognized by their deep red color.

3. The *circumvallate papillæ* are rounded eminences, from eight to ten in number, situated at the base of the tongue, where they form a V-shaped figure. They are quite large, and consist of a central projection of mucous membrane, surrounded by a wall, or circumvallation, from which they derive their name.

The Taste Beakers, supposed to be the true organs of taste, are flask-like bodies, ovoid in form, about the $\frac{1}{300}$ of an inch in length, situated in the epithelial covering of the mucous membrane, on the circumvallate papillæ. They consist of a number of fusiform, narrow cells, and curved so as to form the walls of this flask-like body; in the interior are elongated cells, with large, clear nuclei, the *taste cells*.

Nerves of Taste. The *chorda tympani* nerve, a branch of the facial, after leaving the cavity of the tympanum, joins the 3d division of the 5th nerve between the two pterygoid muscles, and then passes forward in the lingual branches, to be distributed to the mucous membrane of the anterior two-thirds of the tongue. Division or disease of this nerve is followed by a *loss of taste* in the part to which it is distributed.

The *glosso-pharyngeal* enters the tongue at the posterior border of the hyo-glossus muscle, and is distributed to the mucous membrane of the base and sides of the tongue, fauces, etc.

The *lingual branch of the trifacial* nerve endows the tongue with general sensibility; the *hypoglossal* endows it with motion.

The *nerves of taste* in the superficial layer of the mucous membrane form a fine plexus, from which branches pass to the epithelium and penetrate it; others enter the taste beakers, and are directly connected with the taste cells.

The *seat* of the sense of taste has been shown by experiment to be the whole of the mucous membrane over the dorsum of the tongue, soft palate, fauces, and upper part of the pharynx.

The Sense of Taste enables us to distinguish the savor of substances introduced into the mouth, which is different from tactile sensibility. The sapid quality of substances appreciated by the tongue are designated as bitter, sweet, alkaline, sour, salt, etc.

The Essential Conditions for the production of the impressions of taste are (1) a state of solubility of the food; (2) a free secretion of the saliva, and (3) active movements on the part of the tongue, exciting pressure against the roof of the mouth, gums, etc., thus aiding the solution of various articles and their osmosis into the lingual papillæ. Sapid substances, when in a state of solution, pass into the interior of the taste beakers and come into contact, through the medium of the taste cells, with the terminal filaments of the gustatory nerves.

THE SENSE OF SMELL.

The Sense of Smell is located in the mucous membrane lining the upper part of the nasal cavity, in which the olfactory nerves are distributed.

The Nasal Fossæ are two cavities, irregular in shape, separated by the vomer, the perpendicular plate of the ethmoid bone, and the triangular cartilage. They open anteriorly and posteriorly by the anterior and posterior nares, **the** latter communicating with the pharynx. They are lined by mucous membrane, of which the only portion capable of receiving odorous impressions is the part lining the upper one-third of the fossæ.

The Olfactory Nerves, arising by three roots from the posterior and inferior surface of the anterior lobes, pass forward to the cribriform plate of the ethmoid bone, where they each expand into an oblong body, the *olfactory bulb*. From its under surface from fifteen to twenty filaments pass downward through the foramina, to be distributed to the olfactory mucous membrane, where they terminate in long, delicate, spindle-shaped cells, the *olfactory cells*, situated between the ordinary epithelial cells.

The olfactory bulbs are the centres in which odorous impressions are perceived **as** sensations; destruction of these bulbs being attended by an abolition of the sense of smell.

In animals which possess an acute sense of smell, there is a corresponding increase in the development of the olfactory bulbs.

The Essential Conditions for the **sense of** smell are, (1) a special nerve centre capable of receiving impressions and transforming them into odorous sensations. (2) Emanations from bodies which are in a gaseous or vaporous condition. (3) The odorous emanations must be drawn freely through the nasal fossæ; if the odor be very faint, a peculiar inspiratory movement is made, by which the air is forcibly brought into contact with the olfactory filaments. The secretions of the nasal fossæ probably dissolve the odorous particles.

Various substances, as ammonia, horseradish, etc., excite the sensibility of the mucous membrane, which must be distinguished from the perception **of** true odors.

THE SENSE OF SIGHT.

The Eyeball. The eyeball, or organ of vision, is situated within the orbital cavity, and loosely held in position by the fibrous capsule of Tenon. It rests upon a cushion of fat, which never disappears, except in cases of extreme starvation; it is protected from injury by the bony orbital walls, the lids and lashes, and is so situated as to permit an extensive range of vision.

Blood vessels and Nerves. The structures of the eyeball are supplied with blood by the ciliary arteries, which pierce the posterior surface around the optic nerve.

The *Ciliary* or *Ophthalmic ganglion*, about the size of a pin's head, situated in the posterior portion of the orbital cavity, receives filaments of communication from the trifacial or 5th nerve, the motor oculi or 3d nerve, and the sympathetic. From its anterior portion are given off the ciliary nerves, which enter the ball posteriorly and are distributed to the structures of which it is composed.

Structure. The form of the eyeball is that of a sphere; it is about one inch in the transverse diameter, and a little longer in the antero-posterior diameter, on account of its having the segment of a smaller sphere inserted into the anterior surface.

The Sclerotic and Cornea together form the external coat of the eye; the former covering the posterior $\frac{5}{6}$, the latter, the anterior $\frac{1}{6}$. The sclerotic is a dense, opaque, fibrous membrane, varying in thickness from the $\frac{1}{30}$ to the $\frac{1}{25}$ of an inch; it is composed of connective tissue and is slightly vascular. Posteriorly it is continuous with the sheath of the optic nerve, and is pierced by that nerve, as well as by the ciliary vessels and nerves; anteriorly its fibres become quite pale, and after passing into the cornea, transparent. It is a protective covering, and gives attachment to the tendons of the muscles by which the eyeball is moved.

The *Cornea* is a non-vascular, transparent membrane, composed for the most part of connective tissue in which are contained stellate corpuscles filled with a clear fluid. It is covered anteriorly by the basement membrane of the conjunctiva, upon which rests several layers of epithelial cells; posteriorly it is lined by the membrane of Descemet, which is reflected on to the anterior surface of the iris.

The Choroid, the Iris, the Ciliary Muscle and Ciliary Processes, together constitute the middle coat of the eye.

The *Choroid coat*, about the $\frac{1}{50}$ of an inch in thickness, is the vascular tunic; it is connected with the sclerotic by the lamina fusca. From within outward we distinguish the following layers :—

J

1. The lamina supra-choroidea.

2. The elastic layer of Sattler, consisting of two endothelial layers.

3. The chorio-capillaris, choroid proper, or **membrane of** Ruysch, a thick elastic network of arterioles and capillaries lying within **the** outer layer of veins and larger vessels, called the vena vorticosæ.

4. The lamina vitrea, or internal limiting membrane. (The pigmentary layer, formerly classed as belonging to the choroid, is now known to belong, embryologically and physiologically, to the retina.)

The Function of the Choroid is to provide for the vascular supply and drainage of the body of the eye, and to furnish an uniform and high temperature to the retina.

FIG. 15.

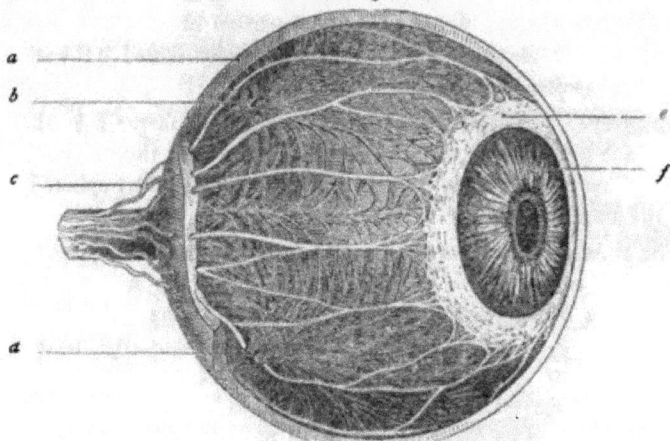

SCLEROTIC COAT REMOVED TO SHOW THE CHOROID, CILIARY MUSCLE AND NERVES.
a. Sclerotic coat. *b.* Veins of the choroid. *c.* Ciliary nerves. *d.* Veins of the choroid.
e. Ciliary muscle. *f.* Iris. (*From Holden's Anatomy.*)

The Iris is a circular, **muscular** diaphragm, placed in the anterior portion of the eye, **and** perforated a little to the nasal side of the centre by a circular opening, **the** pupil; it is attached by its periphery to **the point of** junction of the sclerotic and cornea. It is composed of a connective-tissue stroma, blood vessels and non-striated muscular fibres, circular and radiating. The *circular* fibres surround the margin of the pupil like a sphincter, and are controlled by the 3d pair of nerves; the *radiating* fibres (dilators of the pupil) radiate from the centre toward its circumference, and are controlled by the sympathetic system of nerves.

The *Ciliary muscle* is a grayish circular band, consisting of unstriped muscular fibres, about one-eighth of an inch long, running from before backward; beneath the radiating fibres are small bands of circular fibres running around the eye. It arises from the line of junction of the sclerotic, cornea and iris; passing backward it is attached to the outer surface of the choroid; it is the principal agent in accommodation, and innervated by the 3d pair of nerves.

The Retina forms the internal coat of the eye; in the fresh state it is a delicate, transparent membrane, but soon becomes opaque and of a pinkish tint; it extends forward almost to the ciliary processes, where it terminates in the *ora serrata*. In the posterior portion of the retina, at a point corresponding to the axis of vision, is a rounded, elevated yellow spot, the *limbus luteus*, having a central depression, the *fovea centralis;* about $\frac{1}{10}$ of an inch to the inner side is the point of entrance of the optic nerve, where it spreads out to assist in the formation of the retina. The *arteria centralis retina* pierces the optic nerve near the sclerotic, runs forward in its substance and is distributed in the retina as far forward as the ciliary processes.

The Retina consists of ten distinct layers, from within outward, supported by connective tissue. 1. Membrana limitans interna. 2. Fibres of optic nerve. 3. Layers of ganglionic corpuscles. 4. Molecular layer. 5. Internal granular layer. 6. Molecular layer. 7. External granular layer. 8. Membrana limitans externa. 9. Layer of rods and cones. 10. The layer of pigment cells.

The number of optic nerve fibres in the retina is estimated to be about 800,000, and for each fibre there are about seven cones, 100 rods, and seven pigment cells. The points of the rods and cones are directed toward the choroid, or away from the entering light, and dip into the pigmentary layer. They, with the pigment layer, are the elements intermediating the change of the ethereal vibrations into nervous force. Out of these nerve messages the centre in the occipital lobe fashions the sensations of light, color and form.

In the *Fovea centralis*, at the point of most distinct vision, all of the layers disappear except the layer of rods and cones, which becomes somewhat longer and more slender.

The *Aqueous humor* is a clear fluid, alkaline in reaction, occupying the anterior chamber of the eye; this chamber is bounded in front by the cornea, posteriorly by the iris.

The *Vitreous humor* forms about four-fifths of the entire ball. It supports the retina, and is excavated anteriorly for the reception of the lens; it is transparent, of a jelly-like consistence, and surrounded by a structureless, transparent membrane, the *hyaloid* membrane.

The *Crystalline lens* is situated immediately behind the pupil, in the concavity of the vitreous humor. It is inclosed in a highly elastic, transparent membrane, the *capsule*. The lens is a transparent, double-convex body, ⅓ of an inch transversely, ¼ of an inch antero-posteriorly; it is held in position by the *suspensory ligament*, formed by a splitting of the hyaloid tunic, the external layer of which passes in front of the lens, the internal layer behind it. Its function is to aid in refracting the rays of light and bring them to a focus upon the retina.

FIG. 16.

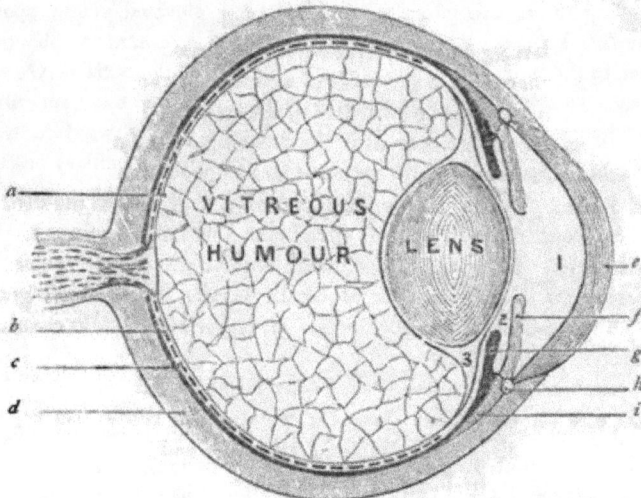

DIAGRAM OF A VERTICAL SECTION OF THE EYE.

1. Anterior chamber filled with aqueous humor. 2. Posterior Chamber. 3. Canal of Petit.
a. Hyaloid membrane. b. Retina (dotted line). c. Choroid coat (black line). d. Sclerotic coat. e. Cornea. f. Iris. g. Ciliary processes. h. Canal of Schlemm or Fontana. i. Ciliary muscle. (*From Holden's Anatomy.*)

Vision. The eye may be regarded as a *camera obscura*, in which images of external objects are thrown upon a screen, the retina, by means of a double convex lens.

The Essential Conditions for proper vision are: 1. *Certain refracting media, e. g.,* cornea, aqueous humor, and crystalline lens, by which the rays of light are so disposed as to form an image. 2. A *diaphragm,* the iris, which, by alternately contracting and dilating, increases or diminishes the amount of light entering the eye. 3. A *sensitive* surface, to receive the image and transmit the luminous impressions through the optic nerve to the brain. 4. A contractile structure, the ciliary muscle, which can so manipulate the lens as to enable external objects to be seen at near or far distances.

The Refracting Apparatus, by which parallel rays of light are brought to a focus on the retina, consists mainly of the crystalline lens, though aided by the cornea and aqueous humor. A ray of light passing through the pupil is refracted and concentrated by the lens at a given point posterior to it. For the correct perception of images of external objects, the rays of light must be accurately focused on the retina; in order that this may be accomplished, the lens must have a certain density, and a proper curvature of its surfaces. When the lens is too convex, its refracting power is greatly increased, the rays of light are brought to a focus in front of the retina, and the visual perception becomes dim and confused. When it is too flat, the rays are not focused at all, and the resulting perception is the same.

The *Crystalline lens*, therefore, produces a distinct perception of the outline and form of external objects.

Action of the Iris. The iris, consisting of contracting and dilating fibres, transmits and regulates the quantity of light passing through its central aperture, the pupil, which is necessary for distinct vision.

If the light be too intense or excessive, the circular fibres contract under the stimulus of the 3d pair of nerves, and the aperture is diminished in size; if the quantity of light be insufficient, the dilating fibres contract under the stimulus of the sympathetic, and the pupillary aperture is increased in size.

The Retina, which is formed partly by the expansion of the optic nerve, and partly by new nervous structures, is the membrane which receives the impressions of light. Its posterior surface, which is in contact with the choroid, and especially the layer of rods and cones, is the sensitive portion, in which the rays of light produce their effects.

The *point* of *most distinct vision* is in the macula lutea, and especially in its central depression, the fovea, which corresponds to the central axis of the eye; it is situated about $\frac{1}{16}$ of an inch to the outside of the entrance of the optic nerve. It is at this point that images of external objects are seen most distinctly, while all around it the perceptions are more or less obscure; at the macula all the layers disappear except the layer of rods and cones.

Blind Spot. At the point of entrance of the optic nerve is a region in which the rays of light make no impression, owing to the absence of the proper retinal elements; the fibres of the optic nerve being insensible to the action of light.

The *course* which a ray of light takes is as follows: After passing through the cornea, lens, and vitreous humor and the layers of the retina, it is finally arrested by the pigmentary layer of the choroid; here it excites in the layer of rods and cones some physical or chemical change, which is then transmitted to the fibres of the optic nerve, and thence to the brain, where it is perceived as a sensation of light.

The Accommodation of the eye to vision for different distances is accomplished by a change in the convexities of the lens, caused by the action of the ciliary muscle. When the eye is accommodated for vision at far distances, the structures are in a passive condition and the lens is flattened; when it is adjusted for vision at short distances, the convexities of the lens are increased.

When the *Ciliary muscle* contracts, the suspensory ligament is relaxed and the lens becomes more convex, in virtue of its own elasticity.

Optical Defects. *Astigmatism* is a condition of the eye which prevents vertical and horizontal lines from being focused at the same time, and is due to a greater curvature of the cornea in one meridian than another.

Spherical aberration is a condition in which there is an indistinctness of an image from the unequal refraction of the rays of light passing through the circumference and the centre of the lens; it is corrected mainly by the iris, which cuts off the marginal rays, and only transmits those passing through the centre.

Chromatic aberration is a condition in which the image is surrounded by a colored margin, from the decomposition of the rays of light into their elementary parts.

Myopia, or *short-sightedness*, is caused by an abnormal increase in the antero-posterior diameter of the eyeball, or by a subnormal refracting power of the lens; it is generally due to the first cause; the lens being too far removed from the retina, forms the image in front of it, and the perception becomes dim and blurred. *Concave* glasses correct this defect, by preventing the rays from converging too soon.

Hypermetropia, or *long-sightedness*, is caused by a shortening of the antero-posterior diameter, or by an excessive refractive power of the lens; the focus of the rays of light would, therefore, be behind the retina. *Convex* glasses correct this defect, by converging the rays of light more anteriorly.

Presbyopia is a loss of the power of accommodation of the eye to near objects, and usually occurs between the ages of 40 and 60; it is remedied by the use of convex glasses.

Accessory Structures. The muscles which move **the eyeball are six** in number; the superior and inferior recti, the external and internal recti, the superior and inferior oblique muscles. The four recti muscles, arising from the apex of the orbit, pass forward and are inserted into the sides of the sclerotic coat; the *superior* and *inferior* muscles rotate the eye around a *horizontal* axis; the *external* and *internal* rotate it around a *vertical* axis.

The *Superior oblique* muscle, having the same origin, passes forward to the inner and upper angle of the orbital cavity, where its tendon passes through a cartilaginous pulley; it is then reflected backward and inserted into the sclerotic just behind the transverse diameter. Its *function* is to rotate the eyeball in such a manner as to direct the pupil *downward* and *outward*.

The *inferior oblique* muscle arises at the inner angle of the orbit and then passes outward and backward, to be inserted into the sclerotic. Its function is to rotate the eyeball and direct the pupil *upward* and *outward*.

By the associated action of all these muscles, the eyeball is capable of performing all the varied and complex movements necessary for distinct vision.

The *Eyelids*, bordered with short, stiff hairs, shade the eye and protect it from injury. On the posterior surface, just beneath the conjunctiva, are the Meibomian glands, which secrete an oily fluid; it covers the edge of the lids, and prevents the tears from flowing over the cheek.

The *Lachrymal Glands* are ovoid in shape, and situated at the **upper** and outer part of the orbital cavity; they open by **from six to eight ducts** at the outer portion of the upper lids.

The *Tears*, secreted by the lachrymal glands, are distributed over the cornea by the lids during the act of winking, and keep it moist and free from dust. The excess of tears passes into the lachrymal ducts, which begin by two minute orifices, one on each lid, at the inner canthus. They conduct the tears into the nasal duct, and so into the nose.

THE SENSE OF HEARING.

The Organ of Hearing is situated in the petrous portion of the temporal bone, and is divided into three portions, viz.: the external ear, the middle ear and the internal ear.

The External Ear consists of two portions, the pinna or auricle, and the external auditory canal. The *former*, consisting of cartilage, which is irregularly folded and covered by integument, is united to the side of the head by ligaments and muscles; the *latter*, partly cartilaginous and partly bony, is about **one** and **a** quarter inches in length; it runs downward and forward from the concha to the middle ear, and is lined by a reflection of the general integument, in which is lodged a number of glands, which secrete the *cerumen*.

The *function* of the external ear is to collect the waves of sound coming from all directions and to transmit them to the membrana tympani.

The Middle Ear or Tympanum is an irregularly-shaped cavity, narrow from side to side, but long in its vertical and antero-posterior diameters.

It is separated from the external ear by the *membrana tympani*, and from the internal ear by a second *membrana tympani;* it communicates posteriorly with the mastoid cells, anteriorly with the pharynx, through the Eustachian tube. It is lined by mucous membrane, and contains three small bones, forming a connected chain running across its cavity.

The *Membrana tympani* is a thin, delicate, translucent membrane, circular in shape and measuring about two-fifths of an inch in diameter; it is received into a delicate ring of bone, which in the adult becomes consolidated with the temporal bone; it is concave externally and situated obliquely, inclining at an angle of 45 degrees.

The membrane consists of three layers: the outer is formed by a reflection of the integument lining the external auditory canal; the middle is composed of fibrous tissue, and the internal of mucous membrane.

The *Function* of the membrana tympani is to receive and transmit the waves of sound to the chain of bones; it is capable of being made tense and lax by the action of the tensor tympani and laxator tympani muscles, so as to vibrate in unison with the waves of sound in the external auditory meatus. When the membrane is relaxed, its vibrations have a greater amplitude, and it appreciates sounds of a low pitch. When it is made tense it vibrates less forcibly and appreciates sounds of a high pitch.

The *Chain of bones* is formed by the malleus, incus and stapes, united together by ligaments. The *malleus* consists of a head, neck and handle, of which the latter is attached to the inner surface of the membrana tympani. The *incus*, or anvil bone, articulates with the head of the malleus by a capsular joint, and with the stapes by the end of its long process. The *stapes* resembles a stirrup in shape; it articulates externally with the long process of the incus, and internally its oval base is applied to the edges of the foramen ovale.

The *Function* of the chain of bones is to transmit the waves of sound across the tympanum to the internal ear; being surrounded by air, and acting as a solid rod, they prevent the vibrations from losing but little in intensity.

The *Tensor tympani muscle* arises mainly from the cartilaginous part of the Eustachian tube; it then passes backward into the tympanic cavity, where it bends at a right angle around a process of bone, and is inserted into the root of the handle of the malleus. Its *function* is to draw the handle of the malleus internally, and thus increase the tension of the membrana tympani, so as to make it capable of vibrating with sounds of greater or less intensity; at the same time it tightens the joints of the chain of bones, so that they may the better conduct waves of sound to the internal ear, with but a slight loss of intensity.

The *Laxator tympani muscle*, arising from the spinous process of the sphenoid bone, passes backward through the Glasserian fissure, into the tympanic cavity, and is inserted into the neck of the malleus. Its *function* is to draw the handle of the malleus outward, and so relax the membrana tympani, and enable it to receive waves of sound of greater amplitude than when it is tense.

The *Stapedius muscle*, emerging from the cavity of the pyramid of bone projecting from the posterior wall of the tympanum, is inserted into the head of the stapes bone. Its *function* is to draw the stapes backward, preventing too great movement of the bone, and at the same time relaxing the membrana tympani.

The *Eustachian tube*, by means of which the middle ear communicates with the pharynx, is partly bony and partly cartilaginous in its structure. It is about one and a half inches in length; commencing at its opening in the pharynx, it passes upward and outward to the spine of the sphenoid bone, where it is slightly contracted; it then gradually dilates as it passes backward into the tympanic cavity. It is lined by mucous membrane, which is continued into the middle ear and into the mastoid cells.

The Eustachian tube permits the passage of air from the pharynx into the middle ear; in this way the pressure of the air within and without the membrana tympani is equalized, which is one of the essential conditions for the reception of sonorous vibrations.

By closing the mouth and nose, and blowing out the cheeks, air can be forced into the middle ear, producing undue pressure and bulging out of the membrana tympani; by making an effort at swallowing, with the mouth and nose closed, the air in the tympanum can be rarefied and the tympanic membrane will be pressed in. In both such cases the acuteness of hearing is very much diminished.

The pharyngeal orifice of the Eustachian tube is opened by the action of certain of the muscles of deglutition, viz.: the levator palati, tensor palati, and at times the palato-pharyngei muscles.

The Internal Ear, or Labyrinth, is located in the petrous portion of the temporal bone, and consists of an osseous and membranous portion.

The Osseous Labyrinth is divisible into three parts, viz.: the vestibule, the semicircular canals and the cochlea.

The *vestibule* is a small, triangular cavity, which communicates with the middle ear by the foramen ovale; in the natural condition it is closed by the base of the stapes bone. The filaments of the auditory nerve enter the vestibule through small foramina in the inner wall, at the fovea hemispherica.

The *Semicircular canals* are three in number: the superior vertical, the inferior vertical and the horizontal, each of which opens into the cavity of the vestibule by two openings, with the exception of the two vertical, which at one extremity open by a common orifice.

The *Cochlea* forms the anterior part of the internal ear. It is a gradually tapering canal, about one and a half inches in length, which winds spirally around a central axis, the *modiolus*, two and a half times. The interior of the cochlea is partly divided into two passages by a thin plate of bone, the *lamina osseous spiralis*, which projects from the central axis two-thirds across the canal. These passages are termed the *scala vestibuli* and the *scala tympani*, from their communication with the vestibule and tympanum. The scala tympani communicates with the middle ear through the *foramen rotundum*, which, in the natural condition, is closed by the second membrana tympani; superiorly they are united by an opening, the helicotrema.

The whole interior of the labyrinth, the vestibule, the semicircular canals, and the scala of the cochlea, contains a clear, limpid fluid, the *perilymph*, secreted by the periosteum lining the osseous walls.

The Membranous Labyrinth corresponds to the osseous labyrinth with respect to form, though somewhat smaller in size.

The *Vestibular* portion consists of two small sacs, the *utricle* and *saccule*. The *Semicircular canals* communicate with the *utricle* in the same manner as the bony canals communicate with the vestibule. The *saccule* communicates with the membranous cochlea by the canalis reuniens. In the interior of the utricle and saccule, at the entrance of the auditory nerve, are small masses of carbonate of lime crystals, constituting the *otoliths*, Their function is unknown.

The *Membranous cochlea* is a closed tube, commencing by a blind extremity at the first turn of the cochlea, and terminating at its apex by a blind extremity also. It is situated between the edge of the *osseous lamina spiralis* and the outer wall of the bony cochlea, and follows it in its turns around the modiolus.

A transverse section of the cochlea shows that it is divided into two portions by the osseous lamina and the basilar membrane : 1. The *scala vestibuli*, bounded by the periosteum and membrane of Reissner. 2. The *scala tympani*, occupying the inferior portion, and bounded above by the septum, composed of the osseous lamina and the membrana basilaris.

The *true membranous canal* is situated between the membrane of Reissner and the basilar membrane. It is triangular in shape, but is partly divided into a triangular portion and a quadrilateral portion by the *tectorial* membrane.

The *Organ of Corti* is situated in the quadrilateral portion of the canal, and consists of pillars of rods, of the consistence of cartilage. They are arranged in two rows ; the one internal, the other external ; these rods rest upon the basilar membrane ; their bases are separated from each other, but their upper extremities are united, forming an arcade. In the internal row it is estimated there are about 3500, and in the external row about 5200 of these rods.

On the inner side of the internal row is a single layer of elongated hair cells ; on the outer surface of the external row are three such layers of hair cells. Nothing definite is known as to their function.

The *Endolymph* occupies the interior of the utricle, saccule, membranous canals, and bathes the strictures in the interior of the membranous cochlea, throughout its entire extent.

The **Auditory Nerve** at the bottom of the internal auditory meatus divides into (1) a vestibular branch, which is distributed to the utricle and semicircular canals ; (2) a cochlear branch, which passes into the central axis at its base, and ascends to its apex ; as it ascends, fibres are given off, which pass between the plates of the osseous lamina, to be ultimately connected with the organ of Corti.

The *Function* of the semicircular canals appears to be to assist in maintaining the equilibrium of the body ; destruction of the vertical canal is followed by an oscillation of the head upward and downward; destruction of the horizontal canal is followed by oscillations from left to right. When the canals are injured on both sides, the animal loses the power of maintaining equilibrium upon making muscular movements.

Function of the Cochlea. It is regarded as possessing the power of appreciating the quality of pitch and the shades of different musical tones. The elements of the organ of Corti are analogous, in some respects, to a musical instrument, and are supposed, by Helmholtz, to be tuned so as to vibrate in unison with the different tones conveyed to the internal ear.

Summary. The waves of sound are gathered together by the pinna and external auditory meatus, and conveyed to the membrana tympani. This membrane, made tense or lax by the action of the tensor tympani and laxator tympani muscles, is enabled to receive sound waves of either a high or low pitch. The vibrations are conducted across the middle ear by the chain of bones to the foramen ovale, and by the column of air of the tympanum to the foramen rotundum, which is closed by the second membrana tympani; the pressure of the air in the tympanum being regulated by the Eustachian tube.

The internal ear finally receives the vibrations, which excite vibrations successively in the perilymph, the walls of the membranous labyrinth, the endolymph, and, lastly, the terminal filaments of the auditory nerve, by which they are conveyed to the brain.

VOICE AND SPEECH.

The Larynx is the organ of voice. Speech is a modification of voice, and is produced by the teeth and the muscles of the lips and tongue, co-ordinated in their action by stimuli derived from the cerebrum.

The Structures entering into the formation of the larynx are mainly the *thyroid, cricoid* and *arytenoid* cartilages; they are so situated and united by means of ligaments and muscles as to form a firm cartilaginous box. The larynx is covered externally by fibrous tissue, and lined internally with mucous membrane.

The Vocal Cords are four ligamentous bands, running antero-posteriorly across the upper portion of the larynx, and are divided into the *two superior* or *false* vocal cords, and the *two inferior* or *true* vocal cords; they are attached anteriorly to the receding angle of the thyroid cartilages and posteriorly to the anterior part of the base of the arytenoid cartilages. The space between the true vocal cords is the *rima glottidis*.

The **Muscles** which have a direct action upon the movements of the vocal cords are nine in number, and take their names from their points of origin and insertion, viz.: the two *crico-thyroid*, two *thyro-arytenoid*, two *posterior crico-arytenoid*, two *lateral crico-arytenoid*, and one *arytenoid* muscles.

The *crico-thyroid* muscles, by their contraction, render the vocal cords *more tense* by drawing down the anterior portion of the thyroid cartilage and approximating it to the cricoid, and at the same time tilting the posterior portion of the cricoid and arytenoid cartilages backward.

The *thyro-arytenoid*, by their contraction, *relax* the vocal cords by **drawing** the arytenoid cartilage forward and the thyroid backward.

The *posterior crico-arytenoid* muscles, by their contraction, rotate the arytenoid cartilages outward and thus separate the vocal cords and enlarge the aperture of the glottis. They principally aid the respiratory movements during inspiration.

The *lateral crico-arytenoid* muscles are antagonistic to the former, and by their contraction rotate the arytenoid cartilages so as to approximate the vocal cords and constrict the glottis.

The *arytenoid* muscle assists in the closure of the aperture of the glottis.

The *inferior laryngeal* nerve animates all the muscles of the larynx, with the exception of the crico-thyroid.

Movements of the Vocal Cords. During respiration the movements of the vocal cords differ from those occurring during the production of voice.

At each *inspiration*, **the** true vocal cords are widely separated, and the aperture **of the glottis is** enlarged by the action of the crico-arytenoid muscles, **which** rotate outward the anterior angle of the base of the arytenoid cartilages; at each *expiration* the larynx becomes passive ; the elasticity of the vocal cords returns them to their original position, and the **air** is forced out by the elasticity of the lungs and the walls of the thorax.

Phonation. As soon as phonation is about to be accomplished a marked change **in the** glottis is noticed with the aid of the laryngoscope. The true vocal cords suddenly become approximated and are made parallel, giving to the glottis the appearance of a narrow slit, the edges of which are capable of vibrating accurately and rapidly ; at the same time their tension is much increased.

With the vocal cords thus prepared, the expiratory muscles force the column of air into the lungs and trachea through the glottis, throwing the edges of the cords into vibration.

The *pitch* of sounds depends upon the exte   to which the vocal cords are made tense and the length of the aperture through which the air passes. In the production of sounds of a high pitch the tension of the vocal cords becomes very marked, and the glottis diminishes in length. When grave sounds, having a low pitch, are emitted from the larynx, **the** vocal cords are less tense and their vibrations are large and loose.

The *quality* of voice depends upon the length, size and thickness of the cords, and the size, form and construction of the trachea, larynx and the resonant cavities of the pharynx, nose and mouth.

The *compass* of the voice comprehends from two to three octaves. The range is different in the two sexes; the lowest note of the male being about one octave lower than the lowest note of the female; while the highest note of the male is an octave less than the highest note of the female.

The *varieties* of voices, *e. g.*, bass, baritone, tenor, contralto, mezzo-soprano and soprano, are due to the length of the vocal cords; being longer when the voice has a low pitch, and shorter when it has a high pitch

Speech is the faculty of expressing ideas by means of combination **of** sounds, in obedience to the dictates of the cerebrum.

Articulate sounds may be divided into *vowels* **and** *consonants*. The *vowel sounds, a, e, i, o, u*, are produced in the larynx by the vocal cords. The *consonantal sounds* are produced in the air passages above the larynx by an interruption of the current of air by the lips, tongue and teeth; the consonants may be divided into: (1) mutes, *b, d, k, p, t, c, g;* (2) dentals, *d, j, s, t, z;* (3) nasals, *m, n, ng;* (4) labials, *b, p, f, v, m;* (5) gutturals, *k, g, c*, and *g* hard; (6) liquids, *l, m, n, r*.

REPRODUCTION.

Reproduction is the function by which the species is preserved, and is accomplished by the organs of generation in the two sexes.

GENERATIVE ORGANS OF THE FEMALE.

The Generative Organs of the Female consist of the ovaries, Fallopian tubes, uterus and vagina.

The Ovaries are two small, ovoid, flattened bodies, measuring one inch and a half in length and three-quarters of an inch in width ; they are situated in the cavity of the pelvis, and imbedded in the posterior layer of the broad ligament ; attached to the uterus by a round ligament, and to the extremities of the Fallopian tubes by the fimbriæ. The ovary consists of an external membrane of fibrous tissue, the *cortical* portion, in which are imbedded the *Graafian vesicles*, and an internal portion, the *stroma*, containing blood vessels.

The Graafian Vesicles are exceedingly numerous, but situated only in the cortical portion. Although the ovary contains the vesicles from the period of birth, it is only at the period of puberty that they attain their full development. From this time onward to the catamenial period, there is a constant growth and maturation of the Graafian vesicles. They consist of an external investment, composed of fibrous tissue and blood vessels, in the interior of which is a layer of cells forming the *membrana granulosa ;* at its lower portion there is an accumulation of cells, the *proligerous disc*, in which the *ovum* is contained. The cavity of the vesicle contains a slightly yellowish, alkaline, albuminous fluid.

The Ovum is a globular body, measuring about the $\frac{1}{125}$ of an inch in diameter; it consists of an external investing membrane, the *vitelline membrane*, a central granular substance, the *vitellus, or yelk*, a nucleus, the *germinal vesicle*, in the interior of which is imbedded the nucleolus, or *germinal spot*.

The Fallopian Tubes are about four inches in length, and extend outward from the upper angles of the uterus, between the folds of the broad ligaments, and terminate in a fringed extremity, which is attached by one of the fringes to the ovary. They consist of three coats : (1) the external, or peritoneal, (2) middle, or muscular, the fibres of which are arranged in a circular or longitudinal direction, (3) internal, or mucous, covered with ciliated epithelial cells, which are always waving from the ovary toward the uterus.

The Uterus is pyriform in shape, and may be divided into a body and neck ; it measures about three inches in length and two inches in breadth in the unimpregnated state. At the lower extremity of the neck is the os externum ; at the junction of the neck with the body is a constriction, the os internum. The cavity of the uterus is triangular in shape, the walls of which are almost in contact.

The walls of the uterus are made up of several layers of non-striated muscular fibres, covered externally by peritoneum, and lined internally by mucous membrane, containing numerous tubular glands, and covered by ciliated epithelial cells.

The Vagina is a membranous canal, from five to six inches in length, situated between the rectum and bladder. It extends obliquely upward from the surface, almost to the brim of the pelvis, and embraces at its upper extremity the neck of the uterus.

Discharge of the Ovum. As the Graafian vesicle matures, it increases in size, from an augmentation of its liquid contents, and approaches the surface of the ovary, where it forms a projection, measuring from one-fourth to one-half an inch in size. The maturation of the vesicle occurs periodically, about every twenty-eight days, and is attended by the phenomena of menstruation. During this period of active congestion of the reproductive organs the Graafian vesicle ruptures, the ovum and liquid contents escape, and are caught by the fimbriated extremity of the Fallopian tube, which has adapted itself to the posterior surface of the ovary. The passage of the ovum through the Fallopian tube into the uterus occupies from ten to fourteen days, and is accomplished by muscular contraction and the action of the ciliated epithelium.

Menstruation is a periodical discharge of blood from the mucous membrane of the uterus, due to a fatty degeneration of the small blood vessels. Under the pressure of an increased amount of blood in the reproductive organs, attending the process of ovulation, the blood vessels rupture, and a hemorrhage takes place into the uterine cavity; thence it passes into the vagina. Menstruation lasts from five to six days, and the amount of blood discharged averages about five ounces.

Corpus Luteum. For some time anterior to the rupture of a Graafian vesicle, it increases in size and becomes vascular ; its walls become thickened, from the deposition of a reddish-yellow, glutinous substance, a product of cell growth from the proper coat of the follicle and the membrana granulosa. After the ovum escapes, there is usually a small effusion of blood into the cavity of the follicle, which soon coagulates, loses its coloring matter, and acquires the characteristics of fibrin, but it takes no part in the formation of the corpus luteum. The walls of the follicle become convoluted, vascular, and undergo hypertrophy, until they occupy the whole of the follicular cavity. At its period of fullest development, the corpus luteum measures three-fourths of an inch in length and half an inch in depth. In a few weeks the mass loses its red color, and becomes yellow, constituting the *corpus luteum*, or *yellow body*. It then begins to retract, and becomes pale; and at the end of two months nothing remains but a small cicatrix upon the surface of the ovary. Such are the changes in the follicle, if the ovum has not been impregnated.

The corpus luteum, after impregnation has taken place, undergoes a much slower development, becomes larger, and continues during the entire period of gestation. The difference between the corpus luteum of the unimpregnated and pregnant condition is expressed in the following table by Dalton :—

	Corpus Luteum of Menstruation.	Corpus Luteum of Pregnancy.
At the end of three weeks.	Three-quarters of an inch in diameter; reddish; convoluted wall	central clot pale.
One month.	Smaller; convoluted wall bright yellow; clot still reddish.	Larger; convoluted wall bright yellow; clot still reddish.
Two months.	Reduced to the condition of an insignificant cicatrix.	Seven-eighths of an inch in diameter; convoluted wall bright yellow; clot perfectly decolorized.
Four months.	Absent or unnoticeable.	Seven-eighths of an inch in diameter; clot pale and fibrinous; convoluted wall dull yellow.
Six months.	Absent.	Still as large as at the end of second month; clot fibrinous; convoluted wall paler.
Nine months.	Absent.	Half an inch in diameter; central clot converted into a radiating cicatrix; external wall tolerably thick and convoluted, but without any bright yellow color.

K

GENERATIVE ORGANS OF THE MALE.

The Generative Organs of the Male consist of the testicles, vasa deferentia, vesiculæ seminales and penis.

The Testicles, the essential organs of reproduction in the male, are two oblong glands, about an inch and a half in length, compressed from side to side, and situated in the cavity of the scrotum.

The proper coat of the testicle, the *tunica albuginea*, is a white, fibrous structure, about the $\frac{1}{25}$ of an inch in thickness; after enveloping the testicle, it is reflected into its interior at the posterior border, and forms a vertical process, the *mediastinum testes*, from which septa are given off, dividing the testicle in lobules.

The substance of the testicle is made up of the *seminiferous tubules*, which exist to the number of 840; they are exceedingly convoluted, and when unraveled are about 30 inches in length. As they pass toward the apices of the lobules they become less convoluted, and terminate in from 20 to 30 straight ducts, the *vasa recta*, which pass upward through the *mediastinum* and constitute the *rete testis*. At the upper part of the mediastinum the tubules unite to form from 9 to 30 small ducts, the *vasa efferentia*, which become convoluted, and form the *globus major* of the *epididymis*; the continuation of the tubes downward behind the testicle and a second convolution constitutes the *body* and *globus minor*.

The seminal tubule consists of a basement membrane lined by granular nucleated epithelium.

The Vas Deferens, the excretory duct of the testicle, is about two feet in length, and may be traced upward from the epididymis to the under surface of the base of the bladder, where it unites with the duct of the vesicula seminalis, to form the ejaculatory duct.

The Vesiculæ Seminales are two lobulated, pyriform bodies, about two inches in length, situated on the under surface of the bladder.

They have an external fibrous coat, a middle muscular coat, and an internal mucous coat, covered by epithelium, which secretes a mucous fluid. The vesiculæ seminales serve as reservoirs, in which the seminal fluid is temporarily stored up.

The Ejaculatory Duct, about ¾ of an inch in length, opens into the urethra, and is formed by the union of the vasa deferentia and the ducts of the vesiculæ seminales.

The Prostate Gland surrounds the posterior extremity of the urethra, and opens into it by from twenty to thirty openings, the orifices of the *pros-*

tatic tubules. The gland secretes a fluid which forms part of the semen, and assists in maintaining the vitality of the spermatozoa.

Semen is a complex fluid, made up of the secretions from the testicles, the vesiculæ seminales, the prostatic and urethral glands. It is grayish-white in color, mucilaginous in consistence, of a characteristic odor, and somewhat heavier than water. From half a drachm to a drachm is ejaculated at each orgasm.

The **Spermatozoa** are peculiar anatomical elements, developed within the seminal tubules, and possess the power of spontaneous movement. The spermatozoa consist of a conoidal head and a long filamentous tail, which is in continuous and active motion; as long as they remain in the **vas** deferens they are quiescent, but when free to move in the fluid of the vesiculæ seminales, become very active.

Origin. The spermatozoa appear at the age of puberty, and are then constantly formed until an advanced age. They are developed from the nuclei of large, round cells contained in the interior of the seminal tubules, as many as fifteen to twenty developing in a single cell.

When the spermatozoa are introduced into the vagina, they pass readily into the uterus and through the Fallopian tubes toward the ovaries, where they remain and retain their vitality for a period of from 8 to 10 days.

Fecundation is the union of the spermatozoa with the ovum during its passage toward the uterus, and usually takes place in the Fallopian tube, just outside of the womb. After floating around the ovum in an active manner, they penetrate the vitelline membrane, pass into the interior of the vitellus, where they lose their vitality, and along with the germinal vesicle entirely disappear.

DEVELOPMENT OF ACCESSORY STRUCTURES.

Segmentation of the Vitellus. After the disappearance of the spermatozoa and the germinal vesicle there remains a transparent, granular, albuminous substance, in the centre of which a new nucleus soon appears; this constitutes the *parent cell*, and is the first stage in the development of the new being.

Following this, the vitellus undergoes *segmentation;* a constriction appears on the opposite sides of the vitellus, which gradually deepens, until the yelk is divided into two segments, each of which has a distinct nucleus and nucleolus; these two segments undergo a further division into four, the four into eight, the eight into others, and so on, until the entire vitellus is divided into a great number of cells, each of which contains a nucleus and nucleolus.

The peripheral cells of this "mulberry mass" then arrange themselves so as to form a membrane, and as they are subjected to mutual pressure, assume a polyhedral shape, which gives to the membrane a mosaic appearance. The central part of the vitellus becomes filled with a clear fluid. A secondary membrane shortly appears within the first, and the two together constitute the external and internal blastodermic membranes.

Germinal Area. At about this period there is an accumulation of cells at a certain spot upon the surface of the blastodermic membranes, which marks the position of the future embryo. This spot, at first circular, soon becomes elongated, and forms the *primitive trace*, around which is a clear space, the *area pellucida*, which is itself surrounded by a darker region, the *area opaca*.

The primitive trace soon disappears, and the area pellucida becomes guitar-shaped; a new groove, the *medullary groove*, is now formed, which develops from before backward, and becomes the neural canal.

Blastodermic Membranes. The embryo, at this period, consists of three layers, viz.: the external and internal blastodermic membranes, and a middle membrane formed by a genesis of cells from their internal surfaces. These layers are known as the epiblast, mesoblast and hypoblast.

The *Epiblast* gives rise to the central nervous system, the epidermis of the skin and its appendages, and the primitive kidneys.

The *Mesoblast* gives rise to the dermis, muscles, bones, nerves, blood vessels, sympathetic nervous system, connective tissue, the urinary and reproductive apparatus and the walls of the alimentary canal.

The *Hypoblast* gives rise to the epithelial lining of the alimentary canal and its glandular appendages, the liver and pancreas, and the epithelium of the respiratory tract.

Dorsal Laminæ. As development advances, the true medullary groove deepens, and there arise two longitudinal elevations of the epiblast, the *dorsal laminæ*, one on either side of the groove, which grow up, arch over and unite so as to form a closed tube, the primitive central nervous system.

The **Chorda Dorsalis** is a cylindrical rod running almost throughout the entire length of the embryo. It is formed by an aggregation of mesoblastic cells, and situated immediately beneath the medullary groove.

Primitive Vertebræ. On either side of the neural canal the cells of the mesoblast undergo a longitudinal thickening, which develops and

extends around the neural canal and the chorda dorsalis, and forms the arches and bodies of the vertebræ. They become divided transversely into four-sided segments.

The *Mesoblast* now separates into two layers ; the external, joining with the epiblast, forms the *somatopleure ;* the internal, joining with the hypoblast, forms the *splanchnopleure ;* the space between them constituting the *pleuro-peritoneal* cavity.

Visceral Laminæ. **The** walls of the pleuro-peritoneal cavity **are** formed **by a** downward prolongation of the somatopleure (the *visceral laminæ*), which, as they extend around in front, pinch off a portion of the yelk sac (formed by the splanchnopleure), which becomes the primitive alimentary canal ; the lower portion, remaining outside of the body cavity, forms the *umbilical vesicle*, which after a time disappears.

Formation of Fœtal Membranes. The Amnion appears shortly after the embryo begins to develop, and is formed by folds of the epiblast and external layer of the mesoblast, rising up in front and behind, and on each side ; these amniotic folds gradually extend over the back of the embryo to a certain point, where they coalesce, and enclose a cavity, the amniotic cavity. The membranous partition between the folds disappears, and the outer layer recedes and becomes blended with the vitelline membrane, constituting the *chorion,* the external covering of the embryo.

The Allantois. As the amnion develops, there grows out from the posterior portion of the alimentary canal a pouch, **or** diverticulum, the *allantois,* which carries blood vessels derived from the intestinal circulation. As it gradually enlarges, it becomes more vascular, and inserts itself between the two layers of the amnion, coming into intimate contact with the external layer. Finally, from increased growth, it completely surrounds the embryo, and its edges become fused together.

In the bird, the allantois is a respiratory organ, absorbing oxygen and exhaling carbonic acid ; it also absorbs nutritious matter from the interior of the egg.

Amniotic Fluid. The amnion, when first formed, is in close contact with the surface of the ovum ; but it soon enlarges, and becomes filled with a clear, transparent fluid, containing albumen, glucose, fatty matters, urea and inorganic salts. It increases in amount up to the latter period of gestation, when it amounts to about two pints. In the space between the amnion and allantois is a gelatinous material, which is encroached upon, and finally disappears as the amnion and allantois come in contact, at about the fifth month.

The Chorion, the external investment of the embryo, is formed by a fusion of the original vitelline membrane, the external layer of the amnion, and the allantois. The external surface now becomes **covered** with villous processes, which increase in number and size by the continual budding and growth of club-shaped processes from the main stem, and **give to the** chorion a shaggy appearance. They consist of a homogeneous granular matter, and are penetrated by branches of the blood vessels derived from the aorta.

The presence of villous processes in the uterine cavity is proof positive of the previous existence of a fœtus. They are characteristic of the chorion, and are found under no other circumstances.

At about the end of the second month the villosities begin to atrophy **and** disappear from the surface of the chorion, with the exception of those situated at the points of entrance of the fœtal blood vessels, which occupy about one-third of its surface, where they continue to grow longer, become more vascular, and ultimately assist in the formation of the placenta ; the remaining two-thirds of the surface loses its villi and blood vessels, and becomes a simple membrane.

The Umbilical Cord connects the fœtus with that portion of the chorion which forms the fœtal side of the placenta. It is a process of the allantois, and contains two arteries and a vein, which have a more or less spiral direction. It appears at the end of the first month, and gradually increases in length, until, at the end of gestation, it measures about twenty inches. The cord is also surrounded by a process of the amnion.

Development of the Decidual Membrane. The interior of the uterus is lined by a thin, delicate, mucous membrane, in which are imbedded immense numbers of tubules, terminating in blind extremities, the *uterine tubules.* At each period of menstruation the mucous membrane becomes thickened and vascular, which condition, however, disappears after the usual menstrual discharge. When the ovum becomes fecundated, the mucous membrane takes on an increased growth, becomes more hypertrophied and vascular, sends up little processes, or elevations from its surface, and constitutes the *decidua vera.*

As the ovum passes from the Fallopian tube into the interior of the uterus, the primitive vitelline membrane, covered with villosities, becomes entangled with the processes of the mucous membrane. A portion of the decidua vera then grows up on all sides, and encloses the ovum, forming the decidua reflexa, while the villous processes of the chorion insert themselves into the uterine tubules, and in the mucous membrane between them.

As development advances the decidua reflexa increases in size, and at about the end of the fourth month comes in contact with the decidua vera, with which it is ultimately fused.

The Placenta. Of all the embryonic structures, the placenta is the most important. It is formed in the third month, and **then** increases in size until the seventh month, when a retrogressive metamorphosis takes place until its separation during labor, at which time it is of an oval or rounded shape, and measures from seven to nine inches in length, six to eight inches in breadth, and weighs from fifteen to twenty ounces. It is most frequently situated at the upper and posterior part of the inner surface of the uterus.

The placenta consists of two portions, a fœtal and a maternal.

The *Fœtal portion* is formed by the villi of the chorion, which, by developing, rapidly increase in size and number. They become branched and penetrate the uterine tubules, which enlarge and receive their many ramifications. The capillary blood vessels in the anterior of the villi also enlarge and freely anastomose with each other.

The *Maternal portion* is formed from that part of the hypertrophied and vascular decidual membrane between the ovum and the uterus, the *decidua serotina*. As the placenta increases in size, the maternal blood vessels around the tubules become more and more numerous, and gradually fuse together, forming great lakes, which constitute sinuses in the walls of the uterus.

As the latter period of gestation approaches, the villi extend deeper into the decidua, while the sinuses in the maternal portion become larger and extend further into the chorion. Finally, from excessive development of the blood vessels, the structures between them disappear, and as their walls come in contact, they fuse together, so that, ultimately, the maternal and fœtal blood are only separated by a thin layer of a homogeneous substance. When **fully** formed, the placenta consists principally of blood vessels interlacing in every direction. The blood of the mother passes from the uterine vessels into the lakes surrounding the villi; the blood from the child flows from the umbilical arteries into the interior of the villi; but there is not at any time an intermingling of blood, the two being separated by a delicate membrane formed by a fusion of the walls of the blood vessels and the walls of the villi and uterine sinuses.

The function of the placenta, besides nutrition, is that of a *respiratory organ*, permitting the oxygen of the maternal blood to pass by osmosis through the delicate placental membrane into the blood of the fœtus; at the **same** time permitting the carbonic acid and other waste products, the result of nutritive changes in the fœtus, to pass into the maternal blood, and so to be carried to the various eliminating organs.

Through the placenta also passes all the nutritious materials of the maternal blood which are essential for the development of the embryo.

At about the middle of gestation there develops beneath the decidual membrane a new mucous membrane, destined to perform the functions of the old when it is extruded from the womb, along with the other embryonic structures, during parturition.

DEVELOPMENT OF THE EMBRYO.

Nervous System. The cerebro-spinal axis is formed within the medullary canal by the development of cells from its inner surfaces, which as they increase fill up the canal, and there remains only the central canal of the cord. The external surface gives rise to the dura mater and pia mater. The neural canal thus formed is a tubular membrane; it terminates posteriorly in an oval dilatation, and anteriorly in a bulbous extremity, which soon becomes partially contracted, and forms the anterior, middle and posterior cerebral vesicles, from which are ultimately developed the cerebrum, the corpora quadrigemina, and medulla oblongata, respectively.

The *anterior vesicle* soon subdivides into two secondary vesicles, the larger of which becomes the hemispheres, the smaller, the optic thalami; **the** *posterior vesicle* also divides into two; the anterior becoming the cerebellum, the posterior, the pons Varolii and medulla oblongata.

About the seventh week the straight chain of cerebral vesicles becomes curved from behind forward and forms three prominent angles. As development advances, the relative size of the encephalic masses changes. The cerebrum developing more rapidly than the posterior portion of the brain, soon grows backward and arches over the optic thalami and the tubercula quadrigemina; the cerebellum overlaps the medulla oblongata.

The surface of the cerebral hemispheres is at first smooth, but at about the fourth month begins to be marked by the future fissures and convolutions.

The Eye is formed by a little bud projecting from the side of the anterior vesicle. It is at first hollow, but becomes lined with nervous matter, forming the *optic nerve* and *retina ;* the remainder of the cavity is occupied by the *vitreous body.* The anterior portion of the pouch becomes invaginated and receives the *crystalline lens*, which is a product of the epiblast, as is also the cornea. The *iris* appears as a circular membrane without a central aperture, about the seventh week; the eyelids are formed between the second and third months.

The **Internal Ear** is developed from the *auditory vesicle*, budding from the third cerebral vesicle; the membranous vestibule appears first, and from it diverticula are given off, which become the semicircular canals and cochlea.

The cavity of the tympanum, the Eustachian tube, and the external auditory canal are the remains of the first branchial cleft; the cavity of this cleft being subdivided into the tympanum and external auditory meatus by the membrana tympani.

The **Skeleton.** The chorda dorsalis, the primitive part of the vertebral column, is a cartilaginous rod situated beneath the medullary groove. It is a temporary structure, and disappears as the true bony vertebræ develop. On either side are the quadrate masses of the mesoblast, the primitive vertebræ, which send processes upward and around the medullary groove, and downward and around the chorda dorsalis, forming in these situations the arches and bodies of the future vertebræ.

More externally the outer layer of the mesoblast and epiblast arch downward and forward, forming the *ventral* laminæ, in which develop the *muscles* and *bones* of the abdominal walls.

The *true cranium* is an anterior development of the vertebral column, and consists of the occipital, parietal and frontal segments, which correspond to the three cerebral vesicles. The *base of the cranium* consists, at this period, of a cartilaginous rod on either side of the anterior extremity of the chorda dorsalis, in which three centres of ossification appear, the *basi-occipital*, the *basi-sphenoidal*, and the *pre-sphenoidal*. They ultimately develop into the basilar process of the occipital bone and the body of the sphenoid.

The *entire skeleton* is at first either membranous or cartilaginous. At the beginning of the second month centres of ossification appear in the jaws and clavicle; as development advances, the ossific points in all the future bones extend, until ossification is completed.

The *limbs* develop from four little buds projecting from the sides of the embryo, which, as they increase in length, separate into the thigh, leg and foot, and the arm, forearm and hand; the extremities of the limbs undergo subdivision, to form the fingers and toes.

Face and Visceral Arches. In the facial and cervical regions the visceral laminæ send up three processes, the *visceral arches*, separated by clefts, the *visceral clefts*.

The *first*, or the *mandibular arches*, unite in the median line to form the lower jaw, and superiorly form the malleus. A process jutting from its base grows forward, unites with the fronto-nasal process growing from above, and forms the upper jaw. When the superior maxillary processes fail to unite, there results the *cleft-palate* deformity; if the integument also fails to unite, there results the *hare-lip* deformity. The space above the mandibular arch becomes the mouth.

The *second arch* develops the incus and stapes bones, the styloid process and ligament, and the lesser cornu of the hyoid bone. The cleft between the first and second arches partially closes up, but there remains an open-ing at the side which becomes the Eustachian tube, tympanic cavity, and external auditory meatus.

The *third arch* develops the body and greater cornu of the hyoid **bone.**

Alimentary Canal and its Appendages. The alimentary canal is formed by a pinching off of the yelk sac by the visceral plates as they grow downward and forward. It consists of three distinct portions, the fore gut, the hind gut, and the central part, which communicates for some time with the yelk sac. It is at first a straight tube, closed at both extremities, lying just beneath the vertebral column. The canal gradually increases in length, and becomes more or less convoluted; at its anterior portion two pouches appear, which become the cardiac and pyloric extremities of the stomach. At about the seventh week the inferior extremity of the intestine is brought into communication with the exterior, by an opening, the anus. Anteriorly the mouth and pharynx are formed by an involution of epiblast, which deepens until it communicates with the fore gut.

The *Liver* appears as a slight protrusion from the sides of the alimen-tary canal, about the end of the first month; it grows very rapidly, attains a large size, and almost fills up the abdominal cavity. The hepatic cells are derived from the intestinal epithelium, the vessels and connective tissue from the mesoblast.

The *Pancreas* is formed by the hypoblastic membrane. It originates in two small ducts budding from the duodenum, which divide and subdivide, and develop the glandular structure.

The *Lungs* are developed from the anterior part of the œsophagus. At first a small bud appears, which, as it lengthens, divides into two branches; secondary and tertiary processes are given off these, which form the bron-chial tubes and air cells. The lungs originally extended into the abdomi-nal cavity, but became confined to the thorax by the development of the diaphragm.

The *Bladder* is formed by a dilatation of that portion of the allantois remaining within the abdominal cavity. It is at first pear-shaped, and communicates with the intestine, but later becomes separated, and opens exteriorly by the urethra. It is attached to the abdominal walls by a rounded cord, the urachus, the remains of a portion of the allantois.

Genito-urinary Apparatus. The *Wolffian bodies* appear about the thirtieth day, as long hollow tubes running along each side of the primitive vertebral column. They are temporary structures, and are sometimes called the primordial kidneys. The Wolffian bodies consist of tubules which run transversely and are lined with epithelium; internally they become invaginated to receive tufts of blood vessels; externally they open into a common excretory duct, the *duct* of the *Wolffian body*, which unites with the duct of the opposite body, and empties into the intestinal canal at a point opposite the allantois. On the outer side of the Wolffian body there appears another duct, the duct of Müller, which also opens into the intestine.

Behind the Wolffian bodies are developed the structures which become either the ovaries or testicles. In the development of the female, the Wolffian bodies and their ducts disappear; the extremities of the Müllerian ducts dilate and form the fimbriated extremity of the Fallopian tubes, while the lower portions coalesce to form the body of the uterus and vagina, which now separate themselves from the intestine.

In the development of the male, the Müllerian ducts atrophy, and the ducts of the Wolffian body ultimately form the epididymis and vas deferens. About the seventh month the testicles begin to descend, and by the ninth month have passed through the abdominal ring into the scrotum.

The *Kidneys* are developed out of the Wolffian bodies. They consist of little pyramidal lobules, composed of tubules which open at the apex into the pelvis. As they pass outward they become convoluted and cup-shaped at their extremities, receive a tuft of blood vessels, and form the Malpighian bodies.

The ureters are developed from the kidneys, and pass downward to be connected with the bladder.

The Circulatory Apparatus assumes three different forms at different periods of life, all having reference to the manner in which the embryo receives nutritious matter and is freed of waste products.

The *Vitelline circulation* appears first and absorbs nutritious material from the vitellus. It is formed by blood vessels which emerge from the body and ramify over a portion of the vitelline membrane, constituting the *area vasculosa*. The heart, lying in the median line, gives off two arches which unite to form the abdominal aorta, from which two large arteries are given off, passing into the vascular area; the venous blood is returned by veins which enter the heart. These vessels are known as the *omphalo-mesenteric arteries and veins*. **The vitelline circulation** is of short dura-tion in the mammals, as the supply of nutritious matter in the vitellus soon becomes exhausted.

The *Placental circulation* becomes established when the blood vessels in the allantois enter the villous processes of the chorion and come into close relationship with the maternal blood vessels. This circulation lasts during the whole of intra-uterine life, but gives way at birth to the adult circulation, the change being made possible by the development of the circulatory apparatus.

The *Heart* appears as a mass of cells coming off from the anterior por-tion of the intestine; its central part liquefies, and pulsations soon begin. The heart is at first tubular, receiving posteriorly the venous trunks and giving off anteriorly the arterial trunks. It soon becomes twisted upon itself, so that the two extremities lie upon the same plane.

The heart now consists of a single auricle and a single ventricle. A septum growing from the apex of the ventricle divides it into two cavities, a right and a left. The auricles also become partly separated by a septum which is perforated by the foramen ovale. The arterial trunk becomes separated by a partition, into two canals, which become, ultimately, the aorta and pulmonary artery. The auricles are separated from the ventri-cles by incomplete septa, through which the blood passes into the ventricles.

Arteries. The aorta arises from the cephalic extremity of the heart and divides into two branches which ascend, one on each side of the intestine, and unite posteriorly to form the main aorta; posteriorly to these first aortic arches four others are developed, so that there are five altogether running along the visceral arches. The two anterior soon disappear. The *third arch* becomes the internal carotid and the external carotid; a part of the *fourth arch*, on the right side, becomes the subclavian artery, and the remainder atrophies and disappears, but on the left side it enlarges and becomes the permanent aorta; the *fifth arch* becomes the pulmonary artery on the left side. The communication between the pulmonary artery and the aorta, the *ductus arteriosus*, disappears at an early period.

Veins. The venous system appears first as two short, transverse veins, the canals of Cuvier, formed by the union of the vertebral veins and the cardinal veins, which empty into the auricle. The inferior vena cava is formed as the kidneys develop, by the union of the renal veins, which, in a short time, receive branches from the lower extremities. The subclavian veins join the jugular as the upper extremities develop. The heart descends in the thorax, and the canals of Cuvier become oblique; they shortly

communicate by a transverse duct, which ultimately becomes the left innominate vein. The left canal of Cuvier atrophies and becomes a fibrous cord. A transverse branch now appears, which carries the blood from the left cardinal vein into the right, and becomes the vena azygos minor; **the** right cardinal vein becomes the vena azygos major.

Circulation of Blood in the Fœtus. The blood returning from the placenta, after having received oxygen, and being freed from carbonic acid, is carried by the umbilical vein to the under surface of the liver; here a portion of it passes through the *ductus venosus* into the ascending vena cava, while the remainder flows through the liver, and passes into the vena **cava by** the hepatic veins. When the blood is emptied into the right **auricle, it** is directed by the Eustachian valve, through the foramen ovale, into the left auricle, thence into the left ventricle, and so into the aorta to all parts of the system. The venous blood returning from the head and upper extremities is emptied, by the superior vena cava, into the **right** auricle, from which it passes into the right ventricle, and thence into **the** pulmonary artery. Owing to the condition of the lung, only **a small** portion flows through the pulmonary capillaries, the greater part passing through the ductus arteriosus, which opens into the aorta at a point below the origin of the carotid and subclavian arteries. The mixed blood now passes down the aorta to supply the lower extremities, but a portion of it is directed, by the hypogastric arteries, to the placenta, to be again oxygenated.

At birth, the placental circulation gives way to the circulation of the adult. As soon as the child begins to breathe, the lungs expand, blood flows freely through the pulmonary capillaries, and the ductus arteriosus begins **to contract.** The foramen ovale closes about the tenth day. The umbilical vein, the ductus venosus, and the hypogastric arteries become impervious in several days, and ultimately form rounded cords.

TABLE OF PHYSIOLOGICAL CONSTANTS.

Mean height of male, 5 feet 6½ inches; of female, 5 feet 2 inches.

Mean weight of male, 145 pounds; of female, 121 pounds.

Number of chemical elements in the human body; from 16 to 18.

Number of proximate principles in the human body; about 100.

Amount of water in the body weighing 145 pounds; 108 pounds.

Amount of solids in the body weighing 145 pounds; 36 pounds.

Amount of food required daily; 16 ounces meat, 19 ounces of bread, 3½ ounces of fat, 52 ounces of water.

Amount of saliva secreted in 24 hours; about 3½ pounds.

 Function of saliva ; converts starch into glucose.

 Active principle of saliva; ptyalin.

Amount of gastric juice secreted in 24 hours; from 8 to 14 pounds.

 Functions of gastric juice; converts albumen into albuminose.

 Active principles of gastric juice ; pepsin and hydrochloric acid.

Duration of digestion; from 3 to 5 hours.

Amount of intestinal juice secreted in 24 hours; about 1 pound.

 Function of intestinal juice ; converts starch into glucose.

Amount of pancreatic juice secreted in 24 hours; about 1½ pounds.

 Active principles of pancreatic juice ; pancreatin and trypsin.

Functions :
 1. Emulsifies fats.
 2. Converts albumen into albuminose.
 3. Converts starch into glucose.

Amount of bile poured into the intestines daily; about 2½ pounds.

Functions :
 1. Assists in the emulsification of fats.
 2. Stimulates the peristaltic movements.
 3. Prevents putrefactive changes in the food.
 4. Promotes the absorption of the fat.

Amount of blood in the body; from 16 to 18 pounds.

Size of red corpuscles ; $\frac{1}{3200}$ of an inch.

Size of white corpuscles ; $\frac{1}{2500}$ of an inch.

Shape of red corpuscles ; circular biconcave disks.

Shape of white corpuscles ; globular.

Number of red corpuscles in a cubic millimetre of blood (the cubic $\frac{1}{25}$ of an inch) ; 5,000,000.

Function of red corpuscles; to carry oxygen from the lungs to the tissues.

Frequency of the heart's pulsations per minute; 72, on the average.

Velocity of the blood movement in the arteries; about 16 inches per second.

Length of time required for the blood to make an entire circuit of the vascular system; about 20 seconds.

Amount of air passing in and out of the lungs at each respiratory act; from 20 to 30 cubic inches.

Amount of air that can be taken into the lungs on a forced inspiration; 110 cubic inches.

Amount of reserve air in the lungs after an ordinary expiration; 100 cubic inches.

Amount of residual air always remaining in the lungs; about 100 cubic inches.

Vital capacity of the lungs; 230 cubic inches.

Entire volume of air passing in and out of the lungs in 24 hours; about 400 cubic feet.

Composition of the air; nitrogen, 79.19, oxygen, 20.81, per 100 parts.

Amount of oxygen absorbed in 24 hours; 18 cubic feet.

Amount of carbonic acid exhaled in 24 hours; 14 cubic feet.

Temperature of the human body at the surface; $98\frac{6}{10}°$ F.

Amount of urine excreted daily; from 40 to 50 ounces.

Amount of urea excreted daily; 512 grains.

Specific gravity of urine; from 1.010 to 1.025.

Number of spinal nerves; 31 pairs.

Number of roots of origin; two; 1st, anterior, motor; 2d, posterior, sensory.

Rate of transmission of nerve force; about 100 feet per second.

Number of cranial nerves; 12 pairs.

Nerves of special sense:
1. Olfactory, or 1st pair.
2. Optic, or 2d pair.
3. Auditory, or 8th pair.
4. Chorda tympani for anterior $\frac{2}{3}$ of tongue.
5. Branches of glosso-pharyngeal, or 8th pair, for posterior $\frac{1}{3}$ of tongue.

Motor nerves to eyeball and accessory structures; motor oculi, or 3d pair; pathetic, or 4th pair; abducens, or 6th pair.

Motor nerves to facial muscles; portio dura, facial, or 7th pair.

Motor nerve to tongue; hypoglossal or 12th pair.

Motor nerve to laryngeal muscles; spinal accessory or 11th pair.

Sensory nerve of the face; trifacial or 5th pair.

Sensory nerve of the pharynx; **glosso-pharyngeal** or 9th pair.

Sensory nerves of **the lungs, stomach, etc.**; pneumogastric or 10th pair.

Length of **spinal cord**; 16 to 18 inches, weight 1½ ounces.

Point of decussation of **motor fibres**; at the medulla oblongata.

Point of decussation of **sensory fibres**; throughout the spinal cord.

Function of **antero-lateral columns** of **spinal cord**; transmit motor impulses from the brain to the muscles.

Functions of the **posterior columns**; assist in the coördination of muscular movements.

Functions of the **medulla oblongata**; controls the functions of insalivation, mastication, deglutition, respiration, circulation, etc.

Functions of the **corpora quadrigemina**; physical centres for sight.

Functions of the **corpora striata**; centres for motion.

Functions of the **optic thalami**; centres for sensation.

Function of the **cerebellum**; centre for the coördination of muscular movement.

Function of the **cerebrum**; centre for intelligence, reason and will.

Centre for **articulate language**; 3d frontal convolution on left side of **cerebrum.**

Number of coats **to the eye; three**; 1st, cornea and sclerotic; 2d, choroid; 3d, retina.

Function of **iris**; regulates the amount of light entering the eye.

Function of **crystalline lens**; refracts the rays of light so as to form an image **on** the retina.

Function of **retina**; receives the impressions of light.

Function of **membrana tympani ; receives** and transmits waves of sound **to** internal ear.

Function of Eustachian **tube ;** regulates the passage of air into and from the middle ear.

Function of **semicircular canals**; assist in maintaining the equipoise of the body.

Function of the **cochlea**; appreciates the shades and combinations of musical tones.

Size of **human ovum** ; $\frac{1}{15}$ of an inch in diameter.

Size of **spermatozoa** ; $\frac{1}{4000}$ of an inch in length.

Function of the **placenta**; acts as a respiratory and digestive organ for the fœtus.

Duration of **pregnancy**; 280 days.

TABLE SHOWING RELATION OF WEIGHTS AND MEASURES OF THE METRIC SYSTEM TO APPROXIMATE WEIGHTS AND MEASURES OF THE U. S.

MEASURES OF LENGTH.

One Myriametre	= 10,000 metres	= 32800. feet.
One Kilometre	= 1,000 "	= 3280. "
One Hectometre	= 100 "	= 328.0 "
One Decametre	= 10 "	= 32.80 "
One Metre	= { the ten millionth part of a quarter of the Meridian of the Earth	= 39.368 inches.
One Decimetre	= the tenth part of one metre	= 3.936 "
One Centimetre	= { the one hundredth part of one metre	= .393 ($\frac{2}{5}$) "
One Millimetre	= { the one thousandth part of one metre	= .039 ($\frac{1}{25}$) "

WEIGHTS.

One Myriagramme	= 10,000 grammes	= 26¾ pounds Troy.
One Kilogramme	= 1,000 "	= 2⅔ " "
One Hectogramme	= 100 "	= 3¼ ounces "
One Decagramme	= 10 "	= 2½ drachms "
One Gramme	= { the weight of a cubic centimetre of water at 4° C.	= 15.434 grains.
One Decigramme	= the tenth part of one gramme	= 1.543 (1½) "
One Centigramme	= { the hundredth part of one gramme	= .154 (⅛) "
One Milligramme	= { the thousandth part of one gramme	= .015 ($\frac{1}{64}$) "

MEASURES OF CAPACITY.

One Myrialitre	= { 10 cubic Metres or the measure of 10 Milliers of water	= 2600. gallons.
One Kilolitre	= { 1 cubic Metre or the measure of 1 Millier of water	= 260. "
One Hectolitre	= { 100 cubic Decimetres or the measure of 1 Quintal of water	= 26. "
One Decalitre	= { 10 cubic Decimetres or the measure of 1 Myriagramme of water	= 2.6 "
One Litre	= { 1 cubic Decimetre or the measure of 1 Kilogramme of water	= 2.1 pints.
One Decilitre	= { 100 cubic Centimetres or the measure of 1 Hectogramme of water	= 3.3 ounces.
One Centilitre	= { 10 cubic Centimetres or the measure of 1 Decagramme of water	= 2.7 drachms.
One Millilitre	= { 1 cubic Centimetre or the measure of 1 gramme of water	= 16.2 minims.

INDEX.

www.ingramcontent.com/pod-product-compliance
Lightning Source LLC
Chambersburg PA
CBHW021806190326
41518CB00007B/479